Elements of Synthesis Planning

T0155763

R.W. Hoffmann

Elements of Synthesis Planning

 Springer

R.W. Hoffmann
Phillips Universität Marburg FB Chemie
Hans-Meerwein-Str.
35042 Marburg
Germany
rwho@chemie.uni-marburg.de

ISBN: 978-3-540-79219-2   e-ISBN: 978-3-540-79220-8

DOI 10.1007/978-3-540-79220-8

Library of Congress Control Number: 2008930867

Cover design: WMXDesign GmbH

Printed on acid-free paper

9 8 7 6 5 4 3 2 1

springer.com

Preface

Synthesis is at the core of organic chemistry. In order for compounds to be studied—be it as drugs, materials, or because of their physical properties—they have to be prepared, often in multistep synthetic sequences. Thus, the target compound is at the outset of synthesis planning.

Synthesis involves creating the target compound from smaller, readily available building blocks. Immediately, questions arise: From which building blocks? In which sequence? By which reactions? Nature creates many highly complex "natural products" via reaction cascades, in which an assortment of starting compounds present within the cell is transformed by specific (for each target structure) combinations of modular enzymes in specific sequences into the target compounds [1, 2]. To mimic this efficiency is the dream of an ideal synthesis [2]. However, we are at present so far from realising such a "one-pot" operation that actual synthesis has to be achieved via a sequence of individual discrete steps. Thus, we are left with the task of planning each synthesis individually in an optimal fashion.

Synthesis planning must be conducted with regard for certain specifications, some of which are due to the structure of the target molecule, and some of which relate to external parameters such as costs, environmental compatibility, or novelty. We will not consider these external aspects in this context. Planning of a synthesis is based on a pool of information regarding chemical reactions that can be executed reliably and in high chemical yield. However, systematic planning of syntheses may in turn identify new types of reactions that would be worthwhile to develop.

For everyone who sets out to synthesize a target compound, planning is the foremost intellectual task. Nevertheless, many steps and sequences in synthesis are chosen by force of habit. Frequently a chemist is not aware of the strengths or weaknesses of a particular projected synthetic sequence. Hence, reflection about the planning of syntheses has a social science component.

One has to pinpoint why chemists react to certain synthesis problems in a distinct way and not otherwise. For this reason, the development of a "logic of chemical synthesis" [3] brought significant progress, which was in due course recognized in 1990 with the Nobel Prize. However, the notion of such a logical approach to synthesis is like a sermon in church: one listens and accepts how one ought to behave, but the next day one's reality reflects a different story.

One can see from the following considerations that the planning of syntheses is in a technical sense not a solved problem: Access to all published synthetic methodology ensures that all chemists are given equal information. Hence, for a given target compound and a given set of parameters, it should be possible to find the single optimal synthesis sequence, or at least a set of equivalent synthetic approaches. All other proposed syntheses should, in theory, be inferior and not worthy of pursuit. A look at today's journals proves this is not the case. The reason for this situation cannot solely be ascribed to the fact that the outcome (e.g., the yields) of individual reactions cannot be predicted with sufficient confidence [4], which causes chemists to scrutinize several reaction variants in order to successfully attain a desired transformation. Rather, the reason for this situation is that the awareness of, and the ability for, retrosynthetic analysis is not equally adopted by every chemist.

However, one should recognize that this imperfection is a cause for intellectual as well as aesthetic excitement in the process of planning a synthesis. The combinatorics of several individual reaction steps leads to such an enormous number of possible (even reasonable) reaction sequences for a given target compound that chemists rely on a subjective choice of projected reactions, which are based on just limited information. Reaction sequences that come to mind at the spur of the moment can appear surprising, impressive, or even elegant. This is where the artistic, creative, and fascinating elements of planning a synthesis arise: to usher a synthesis approach down one particular path and not another. In this sense one can compare the planning of a synthesis with a game of chess [5]. The objectives are clear, yet the number of meaningful moves is so large that any choice becomes subjective. Everyone can reproduce a master chess play, realizing by the third move that he would have chosen another option. At present there exist computer programs for chess, which evaluate all possible options for moves, given a distinct placement of the pieces. The computer will rank these options based on programmed algorithms and will choose accordingly. In the end the game is more rational and logical, but at the same time less exciting.

This analogy demonstrates that "logic-driven, mechanized" considerations of the principles of synthesis planning will demystify this subject. The more rational the plans for syntheses become, the less apparent will be the subjective aspect, the artistic element of the chemist behind the planning process. In this way, the planning of syntheses is being transformed from an art to a technique. Whether one regrets this or not, this is the hallmark of many historic transformations in science. A telltale sign of such a transformation is the development of a specific terminology which is understood solely by the adepts of the field. This is precisely what has happened in the area of synthesis planning during the last three decades.

The author would like to thank Prof. B. H. Lipshutz (UCSB, Santa Barbara, Ca.), as well as Prof. P. S. Baran (Scripps Research Institute, LaJolla, Ca.) and their graduate students for their inspiring discussions and input, which led to the improvement of the present English version beyond the original German version of the book. My thanks also go to C. König and T. Mahnke (Marburg) for checking the contents and references throughout this book.

Marburg, July 2008   *R.W. Hoffmann*

References

1. R. Pieper, C. Kao, C. Khosla, G. Luo, D. E. Cane, *Chem. Soc. Rev.* **1996**, 297–302.
2. P. A. Wender, S. T. Handy, D. L. Wright, *Chem. Ind.* **1997**, 765.
3. E. J. Corey, X.-M. Cheng, *The Logic of Chemical Synthesis*, J. Wiley & Sons, New York, **1989**.
4. J. M. Goodman, I. M. Socorro, *J. Comp.-Aid. Mol. Des.* **2007**, *21*, 351–357.
5. M. H. Todd, *Chem. Soc. Rev.* **2005**, *34*, 247–266.

Contents

Chapter 1
Introduction

Abstract Any synthesis of a target structure requires a plan, which is derived from the target structure by a retrosynthetic analysis. This analysis identifies the bonds to be made in the forward synthesis, i.e., the bond-set. Guiding principles are listed, along which synthesis plans may be developed.

When looking at a target structure, three main aspects should be given attention: the molecular skeleton, the kind and placement of functional groups, and the kind and placement of stereogenic centers. All three aspects impart the planning process for a synthesis; they are interdependent, yet they are not of equal importance. Functional groups may be readily interconverted [1] and, moreover, may be generated from existing $C=O$ and $C=C$ double bond entities. Also, the techniques of stereoselective synthesis have reached a standard [2] such that considerations regarding the generation of stereogenic centers, while an important aspect of synthesis planning, are no longer a paramount problem. In most cases, efficient access to the molecular skeleton remains the major challenge.

Hence, one normally focuses first on the molecular skeleton when planning a synthesis. Consider, for example callystatin A, a target molecule of medium complexity (Scheme 1.1). Try to identify building blocks from which this molecule could be assembled. To do this, one cuts the structure into smaller fragments using *retrosynthetic disconnections*.

There is actually no meaningful alternative in synthesis planning, as S. J. Danishefsky [3] puts it: "It would be improbable, to say the least, to plan the synthesis of a complex target structure through a cognitive process which is fully progressive in nature. Given the stupefying number of ways in which one might begin and proceed, it would seem unlikely that the human mind would go anywhere but in the retrosynthetic direction wherein, at least generally, complexity is reduced as the planning exercise goes on."

Retrosynthetic disconnections are done best in a manner that produces resulting pieces of approximately similar size [4]. Such a tactic will enable the

R.W. Hoffmann, *Elements of Synthesis Planning*,
DOI 10.1007/978-3-540-79220-8_1, © Springer-Verlag Berlin Heidelberg 2009

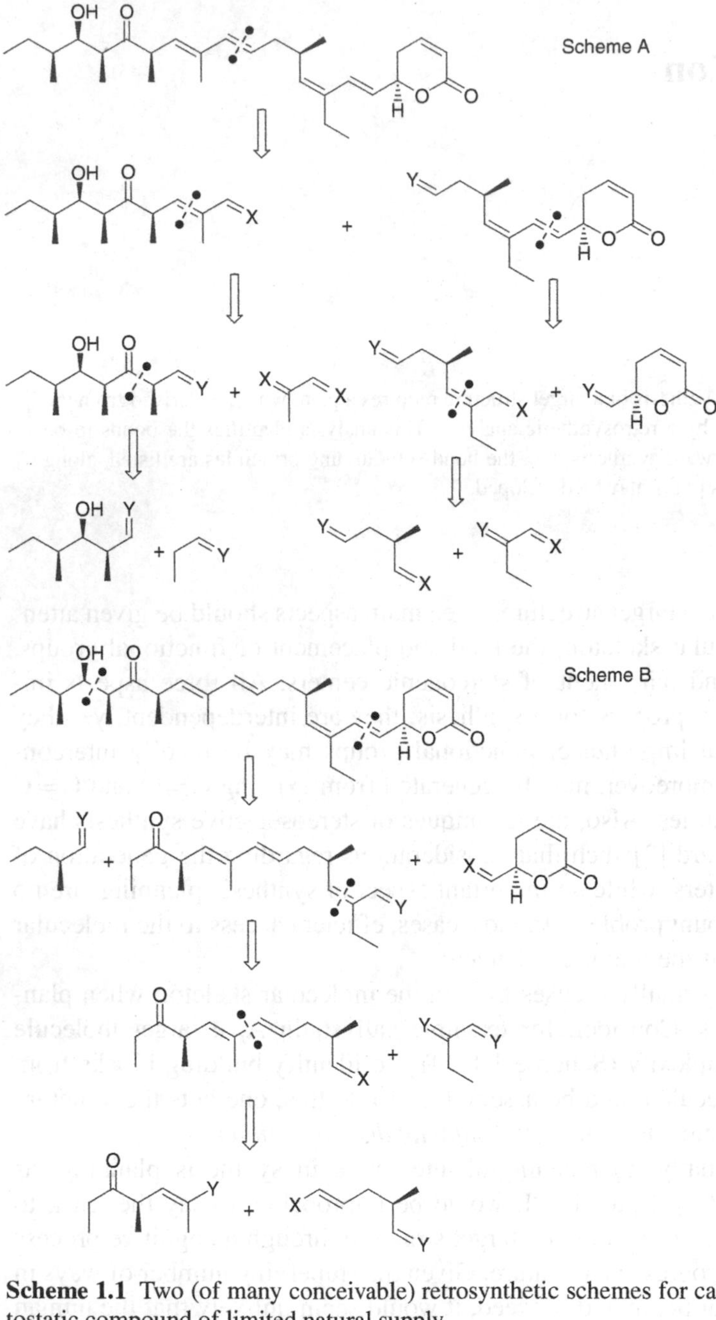

Scheme 1.1 Two (of many conceivable) retrosynthetic schemes for callystatin A, a cytostatic compound of limited natural supply

forward synthesis to proceed in a highly convergent manner (cf. Chap. 8). When one does these cuts based solely on the topology of the target structure, one neglects the knowledge of how to execute bond formation in the forward direction. As the experienced chemist knows which type of bonds he can easily form in actual synthesis, this information, together with the topological considerations, guides the retrosynthetic disconnections. Hence, selecting cuts in retrosynthesis means striking a balance between topological considerations and the availability of easily attainable forward synthetic operations.

At this stage of the planning process, the pieces resulting from the cuts do not have to be fully defined with regard to specific functional groups. Rather, the planning process at this stage yields a highly generalized synthesis plan, as illustrated in Scheme 1.1 for callystatin A [5]. Each conceived retrosynthetic cut is symbolized by a hollow arrow ("retrosynthetic arrow"), whereas each planned forward synthetic step is indicated by a normal arrow. The not yet defined functionalities, enabling bond formation in the synthetic direction, have been designated by the symbols X and Y.

The completion of actual syntheses of callystatin A by these and further routes can be found in reference number [5].

The two retrosynthetic schemes for callystatin A reveal significant differences. In Scheme A, the cuts are done to separate the target into pieces of roughly similar size. By contrast, in Scheme B, the cuts have been done at the periphery of the target structure; thus failing to provide optimal retrosynthetic simplification.

Retrosynthesis schemes generally have the shape of an upside-down tree. At the root is the target structure, while the outer branches constitute the ensemble of starting materials (cf. Scheme 1.2).

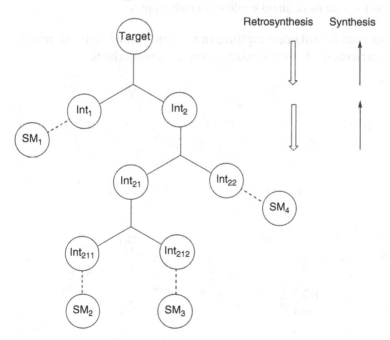

Scheme 1.2 Synthesis tree; Int = intermediate product, SM = starting material

   Inasmuch as the starting materials are only vaguely defined at this level
of the planning process, one stops when the retrosynthetic cuts have yielded
pieces of five to eight skeletal atoms. Building blocks of that size are often commercially available or can be readily obtained by known literature
procedures.

   The formulation of a retrosynthesis scheme is a process in which each step
(cut) is consequential for the possibilities available for the next step (cut).
This is a hierarchical process, because the synthesis tree defines a temporal
sequence in which the bonds are to be formed. Yet obviously there is the
possibility for permutation in the sequences of bond formation.

   The ordering of the synthesis steps and the detailed nature of the starting
materials are left open in a frequently used depiction of the retrosynthetic
analysis. Simply put, bonds that are projected to be formed by synthesis are
marked with a dashed line. This generates a set of marked bonds, which are
referred to as a *bond-set* [4]. Such notation is perfectly suited for comparing
several syntheses of a given target structure (cf. Scheme 1.3).

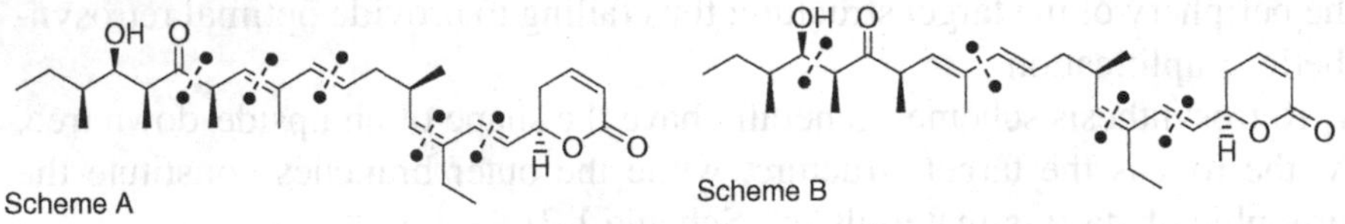

Scheme A   Scheme B

Scheme 1.3 Bond-sets for two realized syntheses of callystatin A

   Bond-set notation is further exemplified in Scheme 1.4, wherein six actual
syntheses of macrolactin A are organized for ready comparison.

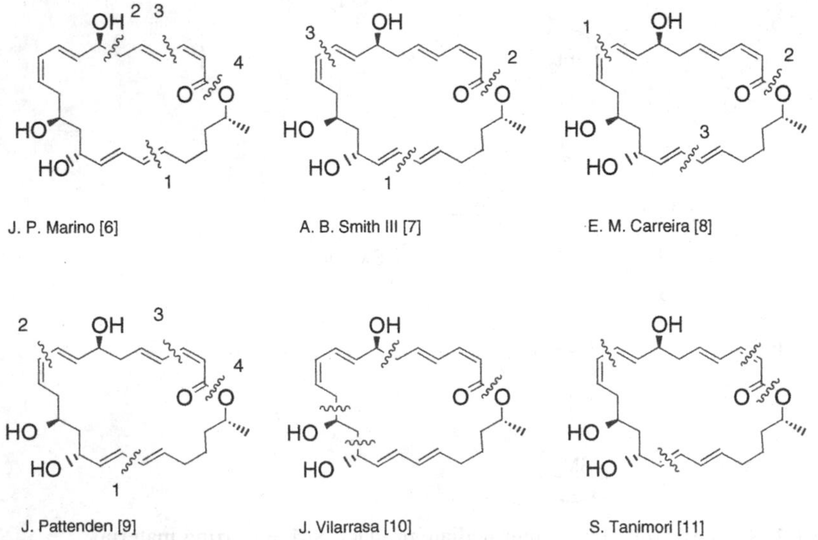

J. P. Marino [6]   A. B. Smith III [7]   E. M. Carreira [8]

J. Pattenden [9]   J. Vilarrasa [10]   S. Tanimori [11]

Scheme 1.4 Realized syntheses (or syntheses in progress) of macrolactin A, an antibi-
otic of limited availability

In almost all cases, the retrosynthetic cuts are performed until pieces of five to eight skeletal atoms result, which are either commercially available or readily synthesized with the appropriate functionality. Comparison of the different bond-sets reveals not only differences in the retrosynthetic approaches, but also common features that originate from particular structural moieties, suggesting certain construction reactions. The expert recognizes the generation of 1,3-diene units by Pd(0)-catalysed coupling reactions. One equally appreciates the options opened by allyl metal addition to aldehydes, by aldol additions, or by carbonyl-olefination reactions. One also notes that olefin metathesis, one of the best methods of forming macrocyclic rings [12], will be problematic in this case, as the target structure contains a plethora of similarly substituted olefinic bonds.

The core of planning a synthesis is to select the individual bonds of a bond-set and the sequence of bond-forming steps in such a manner that in the end an efficient synthesis of the target structure can be realized. Several (quite different) guidelines help in this process. A survey of a multitude of published syntheses reveals that bonds in a bond-set are marked according to:

- the kind and arrangement of functional groups in the target structure = *FG oriented*;
- the peculiarities (branches, rings) of the skeleton of the target structure = *Skeleton oriented*;
- the availability of certain (frequently chiral) building blocks = *Building block oriented*;
- the expertise in certain synthetic methodologies = *Method oriented*.

An optimal synthesis plan rarely follows one of the above options exclusively. Rather, it results from a virtuoso combination of all four of the guidelines. Accordingly, we aspire to learn the basics underlying all of these guidelines and how they relate to the selection of a reasonable bond-set (see Chaps. 2–6). The efficiency of a synthesis not only depends on the bond-sets, but also on the sequence by which the individual bonds are formed (i.e., convergent vs. linear syntheses). In Chap. 8 we will address criteria for rating different synthesis plans. This reveals the sorts of steps that reduce synthetic efficiency (i.e., refunctionalization steps, and the introduction as well as removal of protecting groups or auxiliaries). When the use of protecting groups cannot be avoided completely, there are possibilities by which to minimize the drawbacks of protecting groups, as discussed in Chap. 7.

The points stressed earlier should be highlighted once more: Construction of the skeleton of the target structure is the prime task in synthesis planning, not the placement of functionalities or stereogenic centers. This priority is best reflected when a newly reported target structure arouses the interest of the synthetic community. In such a situation, possible approaches

to the skeleton of the new target structure are evaluated with respect to (func-
tionally) trimmed down versions of the target structure. For example, various inroads to guanacastepene [13] have been explored via construction of
trimmed down guanacastepene congeners [14] (Scheme 1.5).

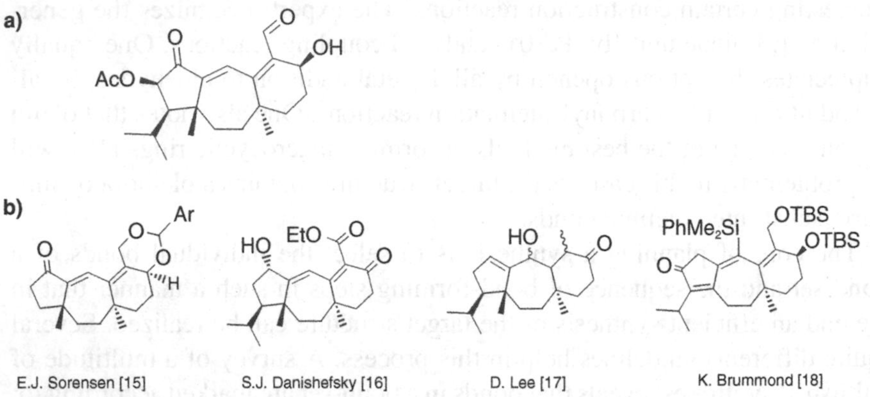

E.J. Sorensen [15]   S.J. Danishefsky [16]   D. Lee [17]   K. Brummond [18]

Scheme 1.5 (a) guanacastepene A, active against antibiotic-resistant bacteria, natural
source no longer accessible; (b) representative trimmed down skeletal versions

The preliminary goals shown in Scheme 1.5b contain the whole or major
parts of the target skeleton (Scheme 1.5a), but lack the complete endowment
of functional groups present in guanacastepene. This was reserved for a second phase of the synthesis effort.

References

1. R. C. Larock, *Comprehensive Organic Transformations*, Wiley-VCH, New York,
   1989.
2. *Houben-Weyl Methods of Organic Chemistry, Vol E21, Stereoselective Synthesis*, (Ed. G. Helmchen, R. W. Hoffmann, J. Mulzer, E. Schaumann) Thieme,
   Stuttgart, **1995**.
3. R. M. Wilson, S. J. Danishefsky, *J. Org. Chem.* **2007**, *72*, 4293–4305.
4. J. B. Hendrickson, *J. Am. Chem. Soc.* **1977**, *99*, 5439–5450.
5. M. Kalesse, M. Christmann, *Synthesis* **2002**, 981–1003.
6. J. P. Marino, M. S. McClure, D. P. Holub, J. V. Comasseto, F. C. Tucci, *J. Am. Chem.
   Soc.* **2002**, *124*, 1664–1668.
7. A. B. Smith III, G. R. Ott, *J. Am. Chem. Soc.* **1998**, *120*, 3935–3948.
8. Y. Kim, R. A. Singer, E. M. Carreira, *Angew. Chem., Int. Ed. Engl.* **1998**, *37*, 1261–
   1263. (*Angew. Chem.* **1998**, *110*, 1321–1323).
9. R. J. Boyce, G. Pattenden, *Tetrahedron Lett.* **1996**, *37*, 3501–3504.
10. A. González, J. Aiguadé, F. Urpí, J. Vilarrasa, *Tetrahedron Lett.* **1996**, *37*,
    8949–8952.

11. S. Tanimori, Y. Morita, M. Tsubota, M. Nakayama, *Synth. Commun.* **1996**, *26*, 559–567.
12. M. E. Maier, *Angew. Chem., Int. Ed.* **2000**, *39*, 2073–2077. (*Angew. Chem.* **2000**, *112*, 2153–2157).
13. M. Mandal, H. Yun, G. B. Dudley, S. Lin, D. S. Tan, S. J. Danishefsky, *J. Org. Chem.* **2005**, *70*, 10619–10637.
14. S. V. Maifeld, D. Lee, *Synlett* **2006**, 1623–1644.
15. W. D. Shipe, E. J. Sorensen, *J. Am. Chem. Soc.* **2006**, *128*, 7025–7035.
16. D. S. Tan, G. B. Dudley, S. J. Danishefsky, *Angew. Chem., Int. Ed.* **2002**, *41*, 2185–2188. (*Angew. Chem.* **2002**, *114*, 2289–2292).
17. T. M. Nguyen, D. Lee, *Tetrahedron Lett.* **2002**, *43*, 4033–4036.
18. K. M. Brummond, D. Gao, *Org. Lett.* **2003**, *5*, 3491–3494.

Chapter 2
Functional Group Oriented Bond-Sets

Abstract During synthesis most skeletal bonds are made by polar bond formation in the vicinity of functional groups. The distance between the bond being formed and the functional group determines the sign of the polarity of the bond forming reaction; distance relationships from 1,1 to 1,3 are considered. When bond formation occurs between two functional groups, a mismatch in polarity may result and has to be corrected by using "umpoled" synthons.

2.1 Polar Bond Formation

When we identify a certain bond in the target structure as one to be made in synthesis (i.e., including it in the bond-set), we should reflect upon the possibilities for constructing such a skeletal bond in the forward synthetic direction. Skeletal bonds are primarily made by polar bond-forming reactions, as illustrated in Scheme 2.1.

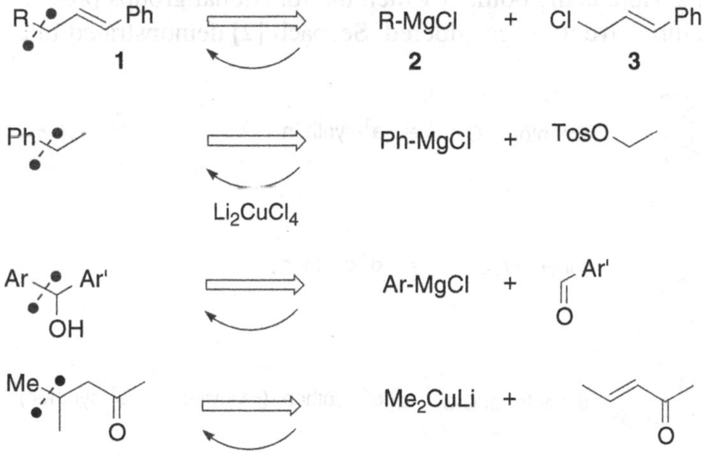

Scheme 2.1 Examples of polar bond-forming reactions

We call the disconnection of compound **1** into potential precursor building blocks **2** and **3** a *retrosynthetic transformation* [1]. In this manner, we capture our knowledge about a synthetic reaction that leads from **2** and **3** to compound **1**. A retrosynthetic transformation is written in the direction opposite to that of a synthetic transformation. As the overwhelming number of synthetic reactions is based on polar bond-forming events, these feature prominently in delineating retrosynthetic transformations of target structures.

During a polar bond-forming reaction, one of the partners (the nucleophile) provides the electron pair that is to form the new bond. The other partner (the electrophile) can, on account of an energetically low-lying empty orbital (LUMO), accommodate the bond-forming electrons. One can choose between two different polarity patterns in order to form a skeletal bond in this manner (Scheme 2.2):

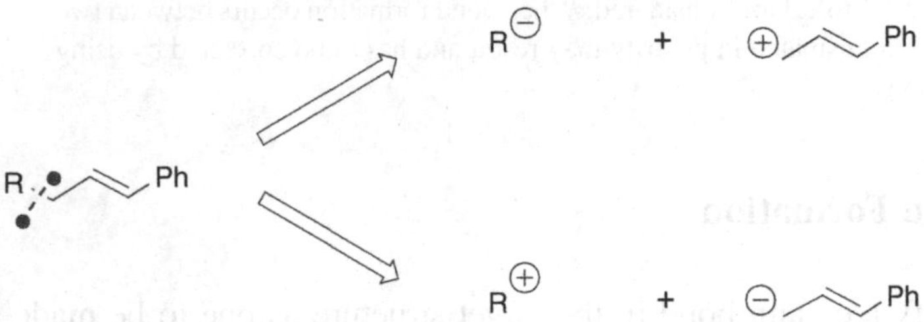

Scheme 2.2 The two different polarity patterns for the formation of a skeletal bond

Which of these options turns out to be more attractive? This depends on how easily a negative or positive (partial) charge can be stabilized in the real synthesis reactants. Here is the point at which the functional groups present in the target structure have to be considered. Seebach [2] demonstrated in a

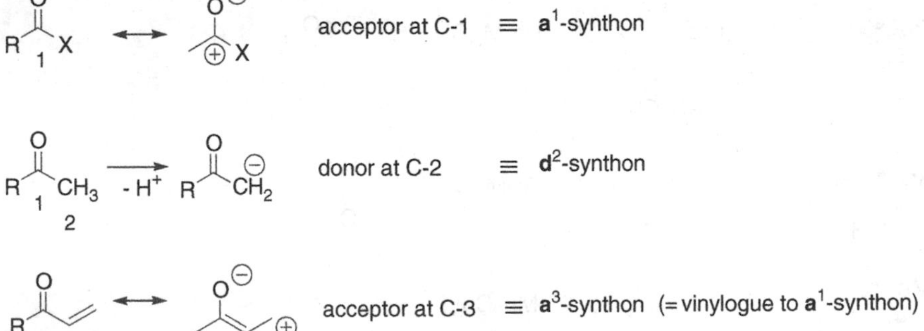

Scheme 2.3 (Partial) charges at or near a carbonyl group; a = acceptor, d = donor. The number designates the position of the reactive center with respect to the skeletal atom of the functional group

fundamental study that a functional group, e.g., a carbonyl group, can help to stabilize either positive or negative charges, depending on the distance from this functional group (Scheme 2.3). In doing this, the removal (or addition) of a proton to accentuate the reactivity pattern—as is done in actual synthetic transformations—is implied.

This leads to clear preferences for the polarity of bond formation at or near a carbonyl group in the target structure (Scheme 2.4):

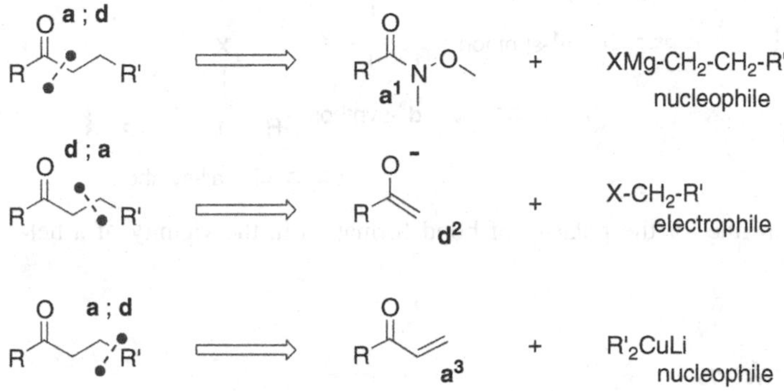

Scheme 2.4 Different polarity in bond-forming reactions depending on the distance from a carbonyl group

This illustrates how a carbonyl group present in the target structure affects the possible types of bond formation in its vicinity. Other functional groups will possess related polarity patterns. To establish such polarity patterns separately for every functional group that commonly occurs in target structures would, however, overcomplicate the exercise, for it would overload the early-planning phase of the synthesis with too many details. Rather, at this stage one relies on the knowledge that most of the important functional groups can be readily (frequently in one-step operations) interconverted [3] (Scheme 2.5).

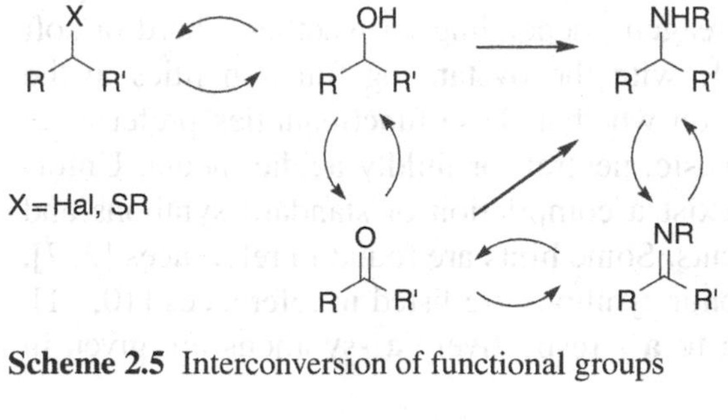

Scheme 2.5 Interconversion of functional groups

As a consequence, one uses a heteroatom substituent X as a generic place-holder for the basic garden varieties of functional groups. This placeholder marks the position of a functional group in the retrosynthetic planning process and determines the preferred polarity for bond formation in the vicinity of such a generalized functional group [4, 5, 6] (Scheme 2.6).

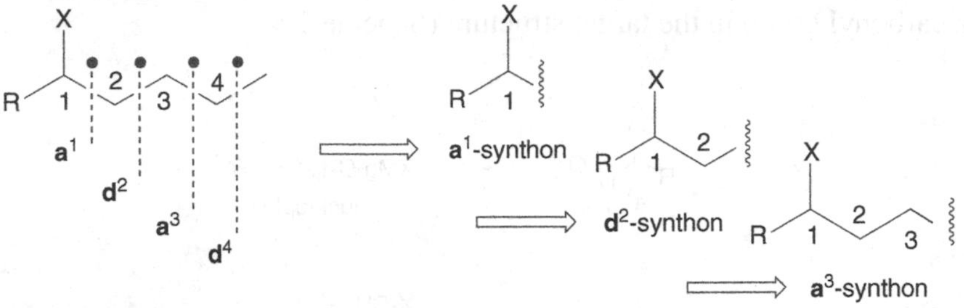

Scheme 2.6 General rule for the polarity of bond formation in the vicinity of a heteroatom substituent

2.1.1 Polar Synthons

In order to develop a general rule for synthesis planning as depicted in Scheme 2.6, we had to make considerable simplifications of the target structure. The relation to the original target structure should, however, be possible at every moment. This implies the ability to connect generalized synthon formulas to existing reactions or reagents.

Retrosynthetic analysis leads to generalized building blocks, which incorporate a reaction principle. These generalized building blocks are called "synthons" [7]. This notation was first employed by Corey [8]. Unfortunately, usage of this term [9] by the chemical community is not consistent. We prefer the usage promoted by Seebach [2], in which a quasi-real axiomatically defined synthon is related to a series of corresponding real reagents. This is illustrated with regard to a d^2-synthon in Scheme 2.7.

In order to carry out the forward synthetic reactions, one has only to choose the most appropriate reagent, depending on whether a hard or soft nucleophile is best compatible with the bystanding functionalities in the intermediates, and depending on whether these functionalities prefer a reaction to be run in strongly basic, neutral, or mildly acidic media. Unfortunately, there does not yet exist a compilation of standard synthons and their corresponding real reagents. Some hints are found in references [2, 7]. Reagents corresponding to donor synthons are listed in references [10, 11]. Some reagents corresponding to a^1- respectively a^3-synthons are given in Scheme 2.8.

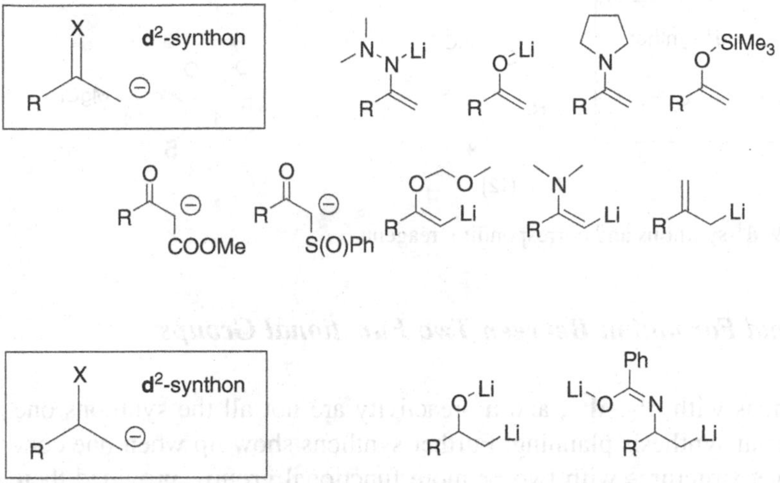

Scheme 2.7 d^2-Synthons and corresponding real reagents

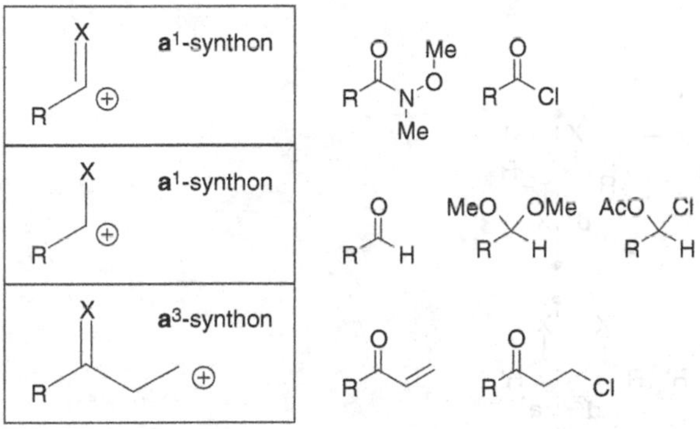

Scheme 2.8 a^1- respectively a^3-synthons and corresponding reagents

Logically, an extension of this sequence would lead to d^4-synthons, for which a few corresponding reagents exist (Scheme 2.9). In practice, compound **4** is likely to display competing d^2- and d^4-reactivity [12]. In the case of compound **5**, there is no relation of the reactivity at C-4 to the functionality at C-1; rather the acetal moiety is masking the carbonyl group, protecting it from the donor reactivity at C-4.

The connection between functionality at C-1 and reactivity at C-n is no longer present. Generally speaking, the synthon concept applies predominantly to distances between the functionality and reactive center of 1–3 skeletal atoms. Thus, the *reach of a functional group* in governing remote reactivity extends no further than the skeletal atom "3".

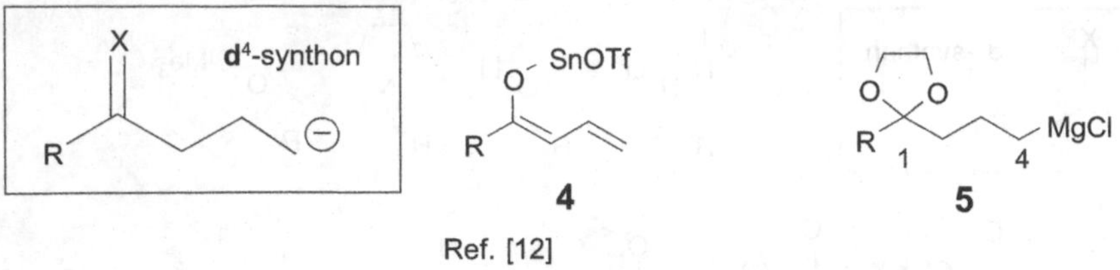

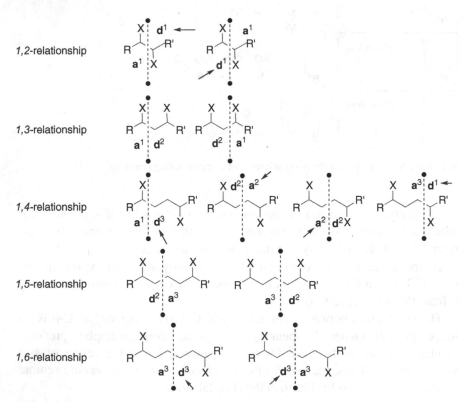

Ref. [12]

Scheme 2.9 d^4-synthons and corresponding reagents

2.1.2 Bond Formation Between Two Functional Groups

The synthons with a^1-, d^2-, and a^3-reactivity are not all the synthons one encounters in synthesis planning. Further synthons show up when one considers target structures with two or more functional groups, provided their distance ranges from 1,2 to 1,6 (counting the skeletal atoms that carry the heteroatom of the functional group). One can select any bond in between these two functional groups for retrosynthetic scission, as summarized in Scheme 2.10.

Scheme 2.10 Bond formation between two functional groups in relationships from *1,2* to *1,6*

Scheme 2.10 demonstrates that:

skeletal bond formation between two functional groups is possible using the hitherto introduced (natural) synthons, as long as the relationship between the functional groups is *1,3* or *1,5*.

skeletal bond formation between two functional groups in a relationship *1,2*, *1,4*, or *1,6* requires in addition a different set of (unnatural) synthons, which are called umpoled synthons [2].

The umpoled synthons are marked in Scheme 2.10 by arrows. The types of synthons that occur in Scheme 2.10 are summarized in Scheme 2.11.

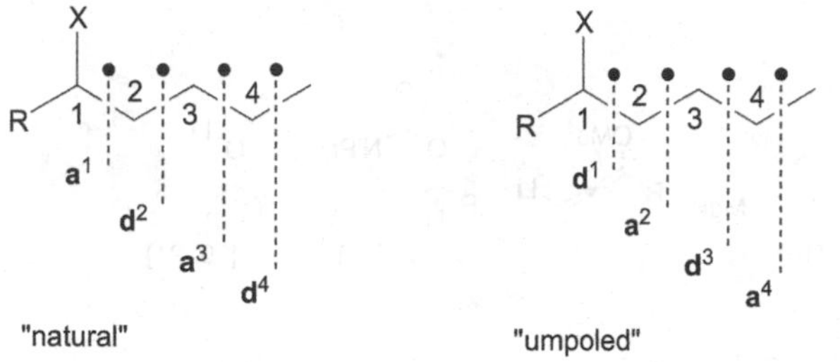

"natural"   "umpoled"

Scheme 2.11 Natural and non-natural (=umpoled) synthons

2.1.3 Umpolung

The conception and development of umpoled synthons were a direct consequence of the above rational concepts for synthesis planning [2]. Before discussing the principles of umpolung and their consequences for planning and efficiency of syntheses, some examples of umpoled synthons are presented in Scheme 2.12.

What is umpolung [2], and how does one use it in synthesis planning? The transformations of an a^1-synthon to a d^1-synthon and the reverse, shown in Scheme 2.13, illustrate this aspect.

Umpolung is a process by which one converts a synthon of natural reactivity into one of "umpoled" or "inverted" reactivity. The accomplishment of this step enables a skeleton bond-forming reaction which, without umpolung, would not have been possible. At the end of the reaction sequence one must reverse the umpolung in order to liberate the functional group with which one started. Thus, the incorporation of an umpoled synthon in a reaction sequence requires at least two additional steps than reaction sequences that

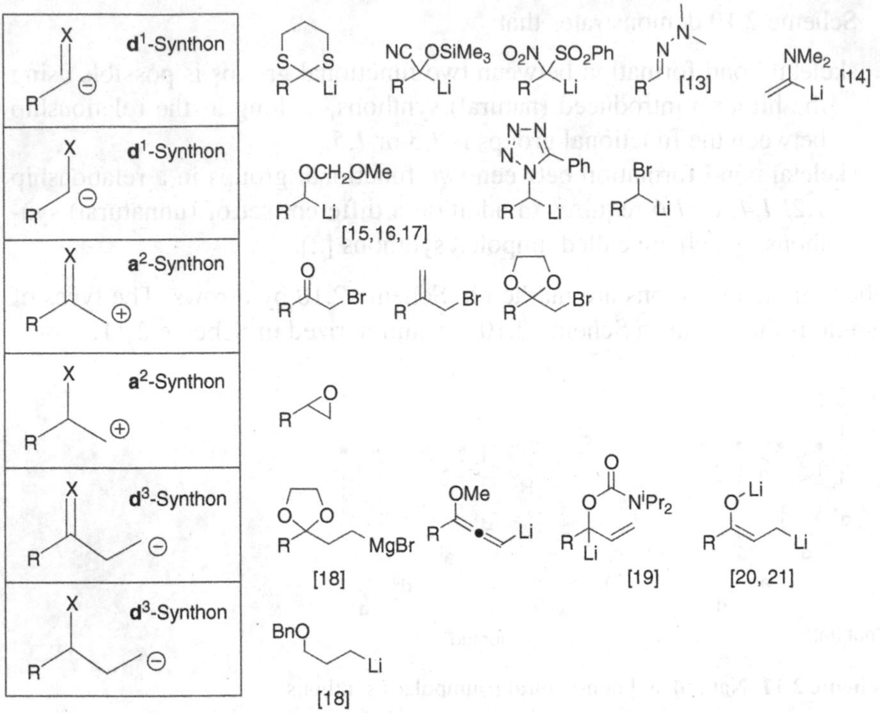

$\mathbf{d^1}$-Synthon

$\mathbf{d^1}$-Synthon

$\mathbf{a^2}$-Synthon

$\mathbf{a^2}$-Synthon

$\mathbf{d^3}$-Synthon

$\mathbf{d^3}$-Synthon

[13] [14]

[15,16,17]

[18] [19] [20, 21]

[18]

Scheme 2.12 $\mathbf{d^1}$-, $\mathbf{a^2}$-, respectively $\mathbf{d^3}$-synthons and corresponding real reagents

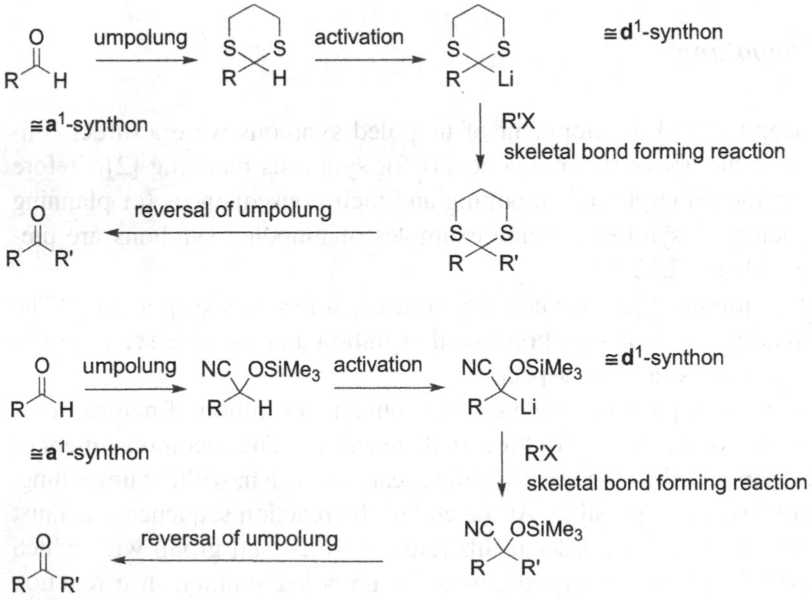

umpolung   activation $\cong \mathbf{d^1}$-synthon

$\cong \mathbf{a^1}$-synthon

R'X
skeletal bond forming reaction

reversal of umpolung

umpolung   activation $\cong \mathbf{d^1}$-synthon

$\cong \mathbf{a^1}$-synthon

R'X
skeletal bond forming reaction

reversal of umpolung

Scheme 2.13 Steps by which umpolung of a reagent is realized rely only on natural synthons. This drawback can be avoided if one succeeds in attaining umpolung in situ by the aid of a catalyst [22].

Nature, in fact, does just that when it converts the a^1-reagent acetaldehyde into a d^1-reagent by thiamine-pyrophosphate. The latter adds to the aldehyde. A subsequent proton shift generates the thiamine conjugate 6, which on account of its enamine unit becomes a d^1-reagent (with reference to the original aldehyde). Nature utilizes this umpolung in situ in a reaction cascade that is continued by skeletal bond formation and reversal of the umpolung to regenerate the aldehyde carbonyl as well as the catalyst (Scheme 2.14).

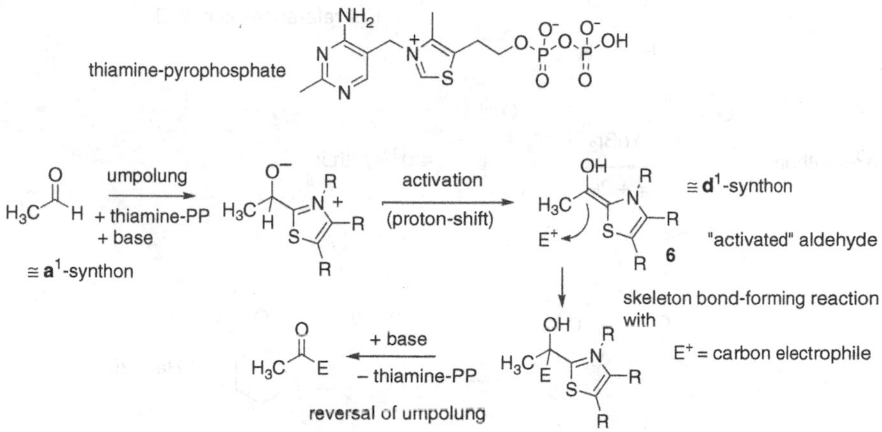

Scheme 2.14 Example of catalytic umpolung in a biosynthetic pathway

The principle ways by which one can attain umpolung have been summarized in a comprehensive paper by Seebach [2]. One far-reaching principle in this context is "*redox-umpolung*." An **a**-synthon may be converted into a **d**-synthon simply by the addition of two electrons (2e-reduction); in reverse, a **d**-synthon is converted into an **a**-synthon by a two-electron oxidation (Scheme 2.15).

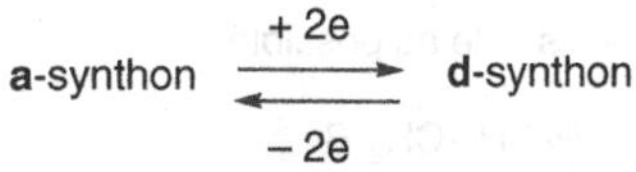

Scheme 2.15 Principle of redox-umpolung

Redox-umpolung can be achieved under actual synthesis conditions. One limitation, though, arises from the fact that the **d**-synthon, for example, is generated via redox-umpolung in the presence of its precursor **a**-synthon.

This opens the possibility for bond formation resulting in a symmetrical coupling product. A standard example is given by pinacol coupling [23, 24] (Scheme 2.16).

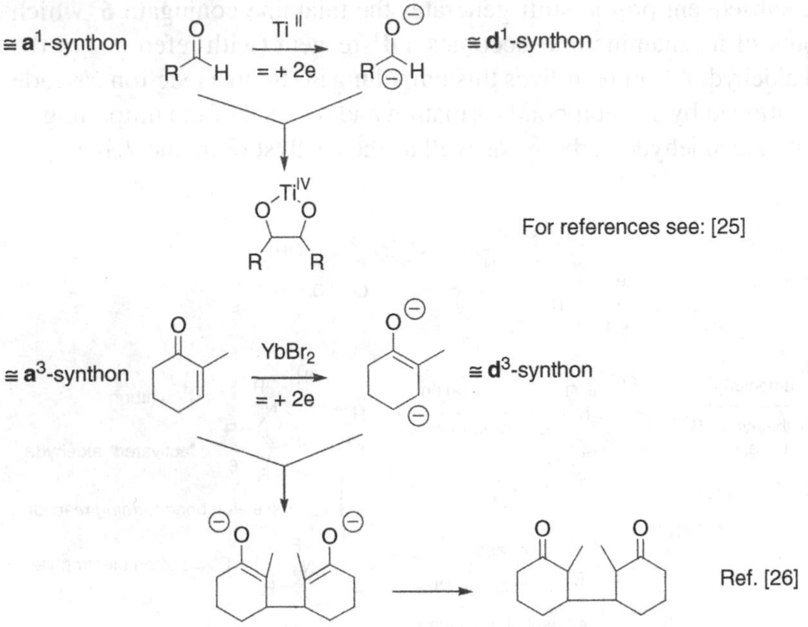

Scheme 2.16 Formation of symmetrical 1,2- and 1,6-difunctionalized skeletons by in situ redox-umpolung

Stoichiometric redox-umpolung becomes possible when the electron transfer process is faster than any coupling step, i.e., when the tendency of both the **a**- and the **d**-synthon for coupling is low. A classical example of such a situation is Grignard formation from alkyl or aryl halides, where the coupling product is generally formed only as a minor side product (Scheme 2.17).

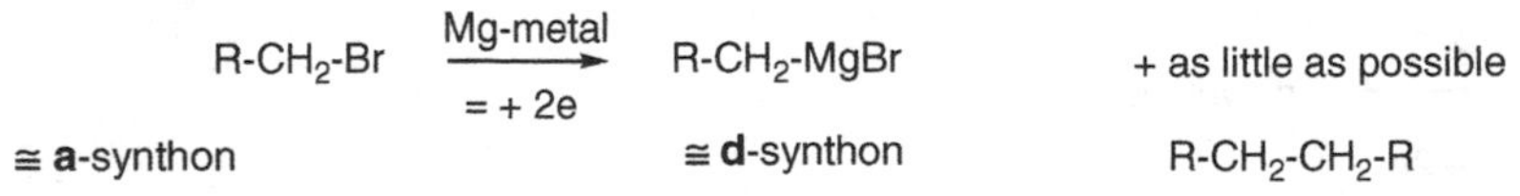

Scheme 2.17 Stoichiometric redox-umpolung during the formation of Grignard reagents

Heterocoupling of two different partners via in situ redox-umpolung is often synthetically more valuable than simple homocoupling. If, for two **a**-synthons, partner A is more readily reduced than partner B, *and* partner B is more reactive towards a **d**-synthon than partner A, then heterocoupling is possible. These conditions are not so likely to occur, yet an example is given in Scheme 2.18. Hence, it seems promising to preassemble the two partners on a transition metal (Cu(I)) first, before the two-electron oxidation is initiated [27].

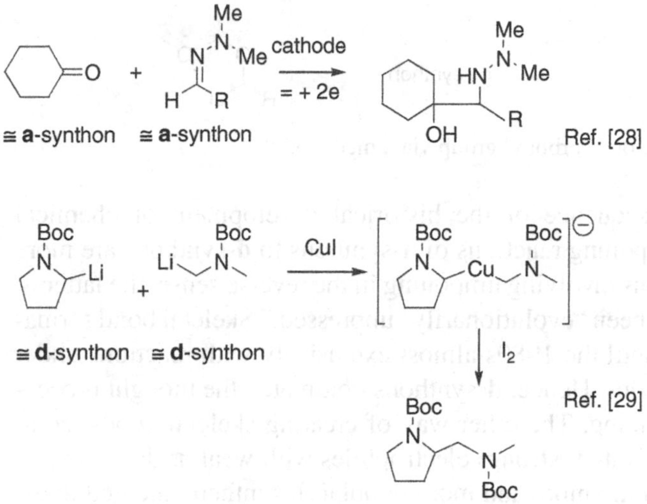

Scheme 2.18 In situ redox-umpolung (direct or copper-mediated) resulting in hetero-coupling products

Intramolecular heterocoupling, i.e., ring closure via in situ redox-umpolung is achieved more readily than the intermolecular version because of favorable effectve concentration of the reacting partners (Scheme 2.19).

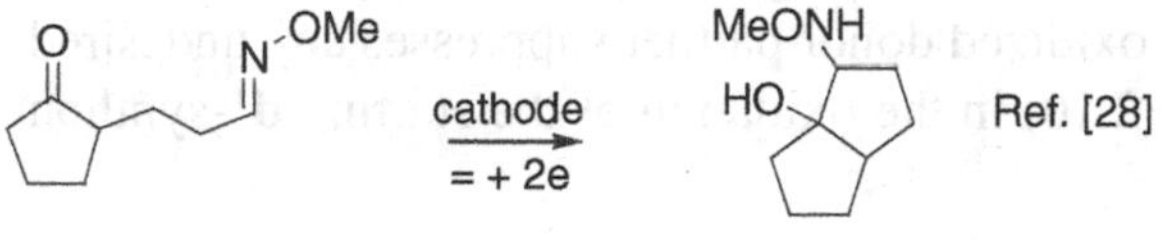

Scheme 2.19 Intramolecular heterocoupling via redox-umpolung

The addition of two electrons during a reductive umpolung of an **a**- to a **d**-synthon may also be achieved indirectly via a mediator such as a metal: as illustrated by the example in Scheme 2.20, in the first step the lithium-tributyltin donates the two electrons necessary for bond formation to the **a**-synthon (the aldehyde). During the subsequent tin/lithium exchange, these two electrons remain at the carbon atom of the former aldehyde, rendering it a nucleophilic center (i.e., a **d**-synthon), and completing the umpolung process [15, cf. also 20].

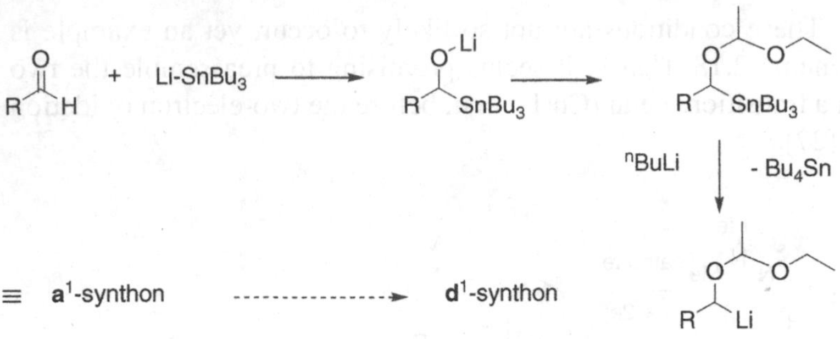

Scheme 2.20 Umpolung of a carbonyl group via a mediator

Perhaps as a consequence of the historical development of chemical methodology, the umpolung reactions of **a**-synthons to **d**-synthons are more prevalent than reactions involving umpolung in the reverse sense, the latter of which appear to have been "evolutionarily suppressed." Skeletal bond formation was conducted until the 1980s almost exclusively with reactions under strongly basic conditions. Hence, **d**-synthons dominated the thought process behind synthesis planning. The other way of creating skeletal bonds, combining Lewis acid activated (strong) electrophiles with weak π-donor nucleophiles, is now becoming more and more popular in synthetic methodology. This requires in due course more unnatural **a**-synthons—that is, umpolung from **d** to **a**.

For skeletal bond-forming reactions in the Lewis-acid realm, the reactive electrophile, an **a**-synthon, is frequently not used in a stoichiometric fashion, but is generated in substoichiometric amounts in situ in the presence of the donor partner. This can be achieved via in situ umpolung from a d^1-synthon to an a^1-synthon, which appears to be a rather roundabout way to reach a natural synthon. The in situ technique, though, has an advantage, because the presence of the less readily oxidized donor partner suppresses any undesired homocoupling (cf. Scheme 2.16) in the oxidation of the starting d^1-synthon (Scheme 2.21).

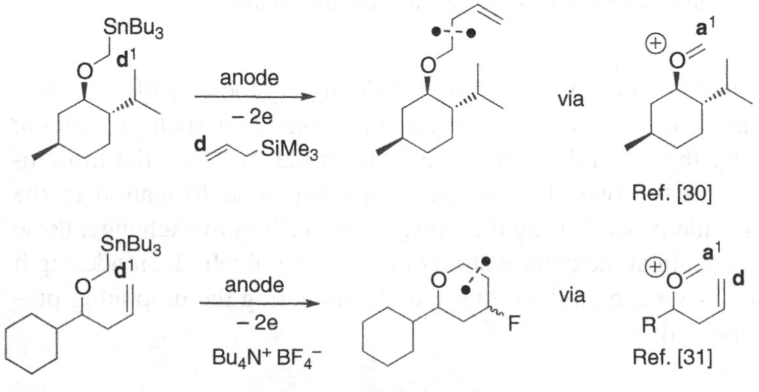

Scheme 2.21 Anodic (oxidative) in situ umpolung from d^1 to a^1

Attaining such an oxidative umpolung is not restricted to electrochemistry [32]; rather, any oxidizing agent may be applied. This is illustrated in Scheme 2.22 with the generation of a d^1-building block, followed by in situ oxidative umpolung to an a^1-building block.

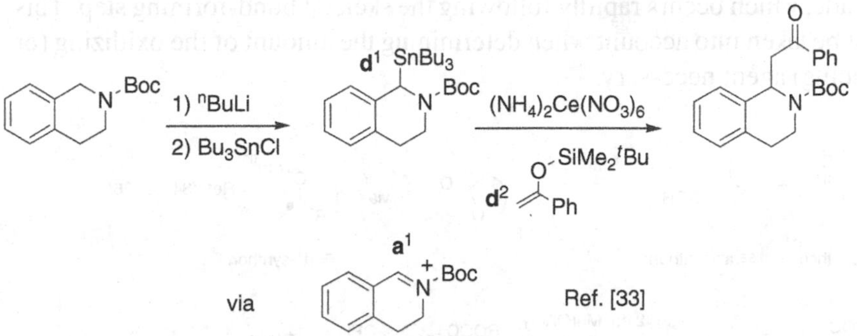

Scheme 2.22 Oxidative in situ umpolung from d^1 to a^1

The following (Scheme 2.23) generation of an a^2-synthon from a d^2-synthon is so elementary that one realizes only by hindsight that this is an oxidative umpolung process.

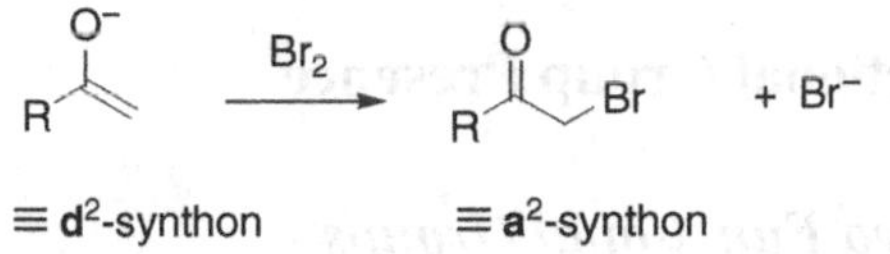

$\equiv d^2$-synthon $\equiv a^2$-synthon

Scheme 2.23 Oxidative umpolung from d^2 to a^2

One generally associates with redox-umpolung the removal or addition of *two* electrons, which results in the correct stoichiometry of the reagents to be employed. It is less well recognized, however, that removal or addition of just *one* electron is sufficient to cause umpolung in reactivity (Scheme 2.24).

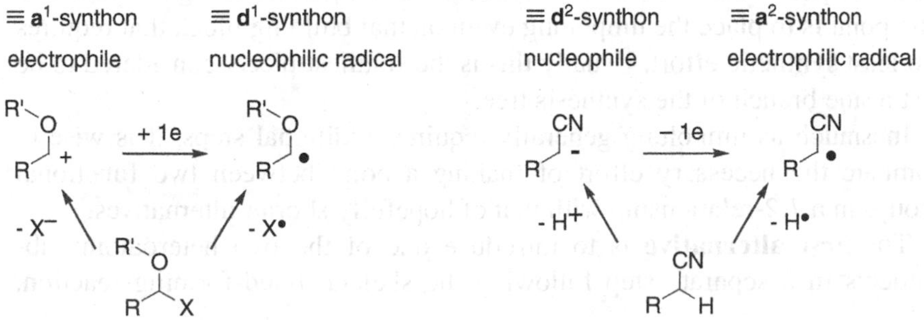

$\equiv a^1$-synthon	$\equiv d^1$-synthon	$\equiv d^2$-synthon	$\equiv a^2$-synthon
electrophile	nucleophilic radical	nucleophile	electrophilic radical

Scheme 2.24 One-electron umpolung

Such one-electron umpolung is frequently advantageous in synthesis, as it may be achieved in situ, thus obviating the two extra steps required by stepwise umpolung tactics. The examples in Scheme 2.25 illustrate applications of one-electron umpolung processes. To satisfy stoichiometry of the overall reaction, though, a second electron has to be transferred during the reaction cascade, which occurs rapidly following the skeletal bond-forming step. This must be taken into account when determining the amount of the oxidizing (or reducing) agent necessary.

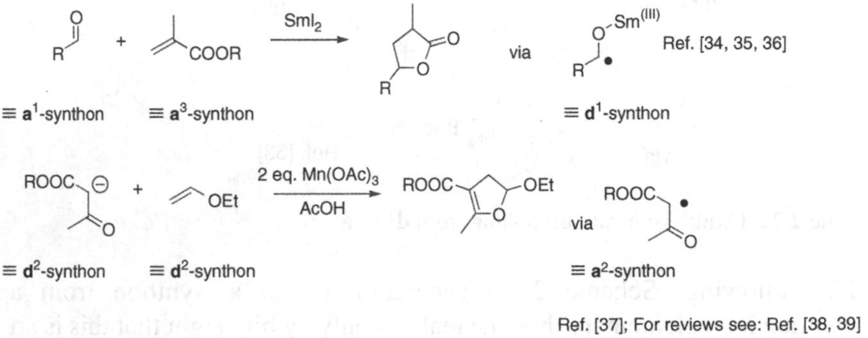

Scheme 2.25 In situ one-electron redox umpolung

2.2 Bond-Sets According to Functional Group Presence

2.2.1 1,2-Relationship Between Two Functional Groups

Our considerations about umpolung originated from the notion of a target structure having two functional groups in a *1,2-* or *1,4-*relationship. Forming a bond between these two functionalities required the use of umpoled synthons. During the forward synthesis, umpolung necessitates at least one, but in most cases two, additional steps. In retrosynthetic analysis, one tries to minimize this drawback by placing the umpolung event on the smaller (reagent) part rather than on the larger (substrate) part resulting from the cut. The point is to place the umpolung event on that building block that requires the least synthetic effort. Usually this is the smallest piece, considered to be just a side branch of the synthesis tree.

Inasmuch as umpolung generally requires additional steps, it is wise to compare the necessary effort of making a bond between two functional groups in a *1,2-*relationship with that of hopefully shorter alternatives.

The **first alternative** is to introduce one of the two heteroatom substituents in a separate step following the skeletal bond-forming reaction.

This too requires an extra step, but just one! In the generic example given in Scheme 2.26, this could be an electrophilic oxygenation [40] or amination [41, 42] of an enolate.

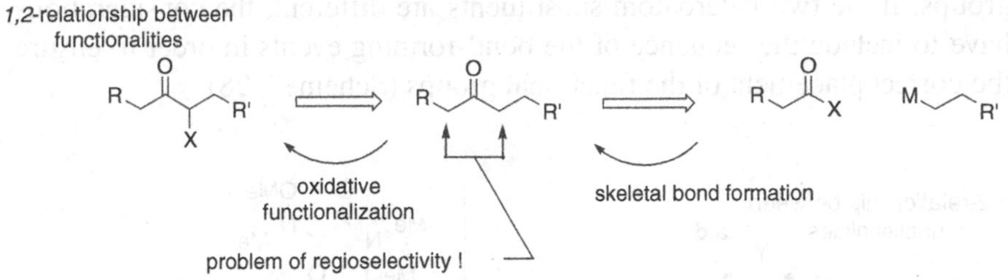

Scheme 2.26 Subsequent oxidative introduction of a second heteroatom

In choosing such a tactic, the position of the second heteroatom is not already fixed in the skeletal bond-forming step. A *regioselectivity* problem may arise in the installment of the second functionality in a distinct position relative to the first functionality. If no obvious solution exits, this alternative approach has to be rejected.

When one has to consider an additional refunctionalization step anyhow—be it during umpolung or upon introduction of the second functionality—a **second alternative** comes up, one in which *both* heteroatom substituents are introduced after the skeletal bond-forming step. For a pair of functionalities in a *1,2*-relationship, this suggests the use of a carbon-carbon double bond as a pro-functionality, as in Scheme 2.27.

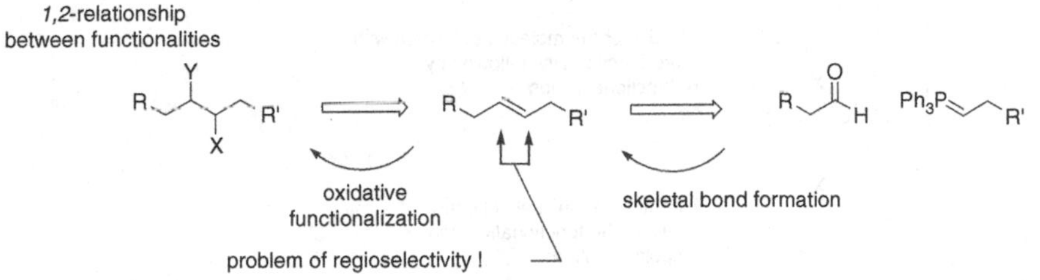

Scheme 2.27 Subsequent oxidative introduction of both heteroatom substituents

Using this alternative, one first assembles the molecular skeleton with a carbon-carbon double bond at the location that will eventually bear the heteroatom substituents. The latter will be introduced in a second step by oxidative addition. In case the two heteroatom substituents are different, a solution for the regioselectivity problem must be sought.

The **third alternative** is to renounce a bond formation between the two functionalities (here in a *1,2*-relationship) by changing to a *building block* *oriented retrosynthesis.* One checks the availability of suitable building blocks which carry two heteroatom substituents in a *1,2-*relationship and which could be incorporated into the target structure. This means that the bonds to be made lie outside of the segment carrying the two functional groups. If the two heteroatom substituents are different, the considerations have to include the sequence of the bond-forming events in order to ensure the correct placement of the functional groups (Scheme 2.28).

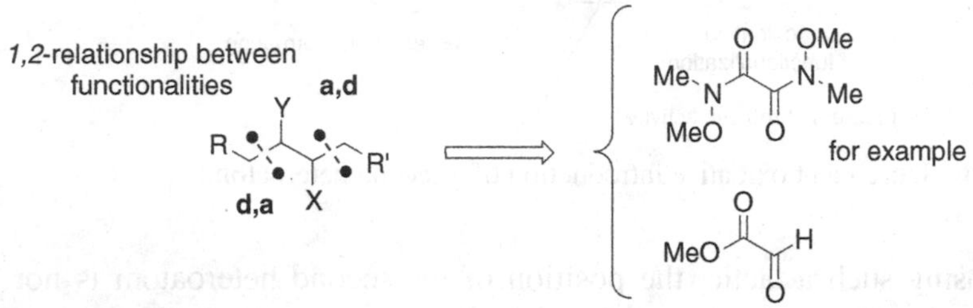

Scheme 2.28 Building block oriented approach to *1,2*-di-heterosubstituted skeletons

The preferred ways to arrive at a *1,2*-di-heterosubstituted skeleton are summarized in Scheme 2.29.

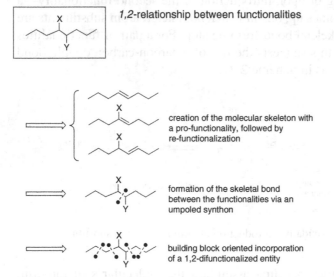

Scheme 2.29 Options for the construction of *1,2*-dihetero-substituted skeletons

Let us conclude this section with a discussion of the synthesis of dihydropalustramic acid (**7**) (Scheme 2.30). In this case, all four diastereomers of the target structure were desired. Thus, we may disregard any aspects of stereochemistry during preliminary retrosynthetic considerations. Looking at compound **7**, one identifies a main chain of the skeleton carrying a *1,2-* and a *1,3*-relationship of heteroatom substituents. The piperidine ring can be depicted as a *1,5*-relationship of heteroatoms along the main carbon chain.

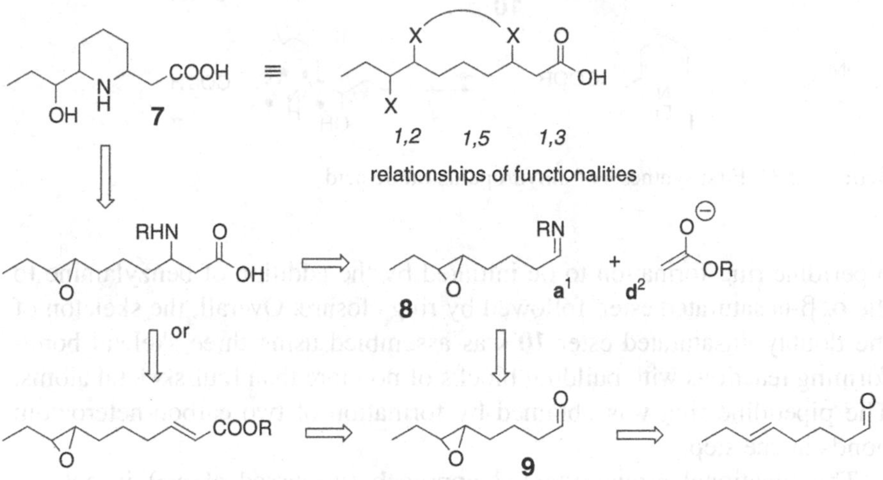

relationships of functionalities

Scheme 2.30 Retrosynthesis of dihydropalustramic acid

It appears advantageous to approach the 1,2-difunctionality from the double bond profunctionality via an epoxide. The 1,3-difunctionality can be obtained with bond formation using natural synthons. For instance, one could employ a Mannich reaction between an ester enolate and an imine **8**, which could arise from an aldehyde in a "functional group interchange" (FGI) step. This two-step procedure to address the *1,3*-di-heteroatom relationship compares with an alternative two-step approach: begin with aldehyde **9** and extend the skeleton by forming an α,β-unsaturated ester. The 1,3-di-heterofunctionality can now be generated by nucleophilic addition of ammonia or an amine to the α,β-unsaturated ester. This sets the stage for an ensuing nucleophilic opening of the epoxide moiety. Because of the intramolecular nature of the attack on the epoxide, the piperidine ring is formed regioselectively (as predicted by Baldwin rules! [43]).

The initial syntheses of dihydropalustramic acid (Eugster [44, 45]) (Scheme 2.31) contain several elements of the synthesis planning depicted in Scheme 2.30 above.

During the synthesis in Scheme 2.31 the molecular skeleton was assembled first and subsequently decorated with the appropriate functional groups. To obtain this the ester **10** contained two double bonds as profunctionality. Selective epoxidation of the more electron-rich double bond enabled the

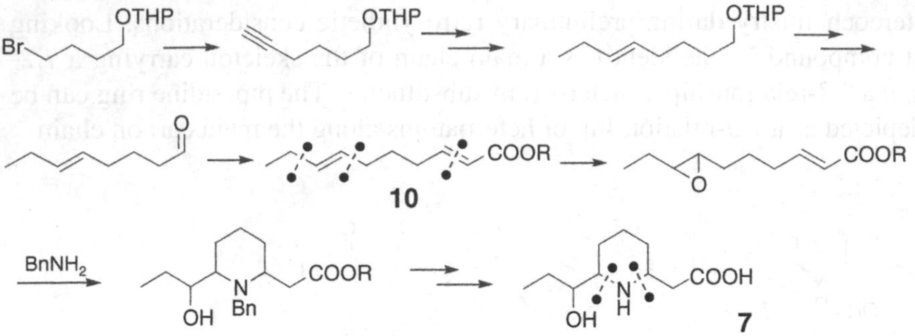

OTHP
Br
OTHP
OTHP
O
COOR
10
COOR
O
BnNH2
COOR
N
OH
Bn
COOH
N
H
OH
7

Scheme 2.31 First synthesis of dihydropalustramic acid

piperidine ring formation to be initiated by the addition of benzylamine to the α, β-unsaturated ester, followed by ring closure. Overall, the skeleton of the doubly unsaturated ester **10** was assembled using three skeletal bond-forming reactions with building blocks of no more than four skeletal atoms. The piperidine ring was obtained by formation of two carbon-heteroatom bonds in one step.

The functional group oriented approach (discussed above) is not the only means by which dihydropalustramic acid has been synthesized. Other creative approaches, the key steps of which are illustrated in Scheme 2.32, open more versatile pathways to this target.

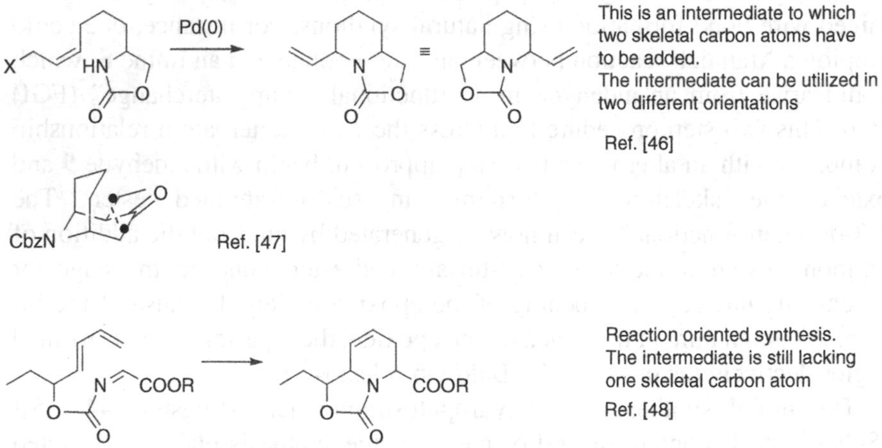

Pd(0)
X
HN
O
O
N
O
O
≡
N
O
O
This is an intermediate to which two skeletal carbon atoms have to be added.
The intermediate can be utilized in two different orientations
Ref. [46]
CbzN
O
Ref. [47]
N
COOR
O
O
N
COOR
O
O
Reaction oriented synthesis.
The intermediate is still lacking one skeletal carbon atom
Ref. [48]

Scheme 2.32 Further synthetic approaches to dihydropalustramic acid

2.2.2 1,4-Relationship Between Two Functional Groups

A major part of the synthetic methods useful for addressing the presence of oxygen functionalities in a *1,4*-relationship on a carbon skeleton has been developed during efforts to synthesize pyrenophorin and vermiculin (Scheme 2.33). It is thus convenient to examine 1,4-difunctional relationships in the context of these two natural products [49]. These macrodilactones are comprized of two identical hydroxy acids. We will again refrain from discussing the stereochemical issues in order to focus on the fundamental aspects of the synthesis of the hydroxy acid subunit.

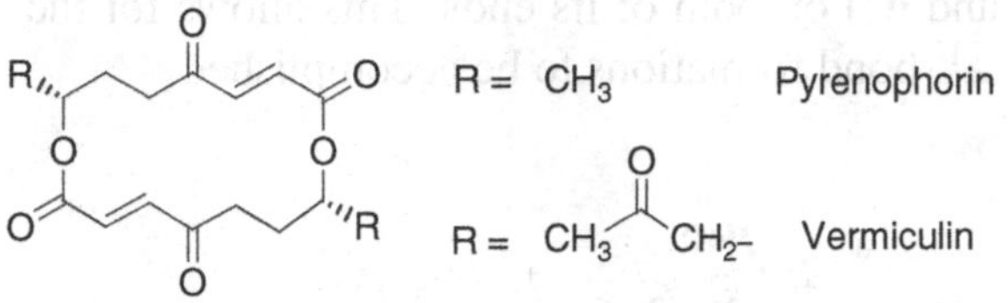

Scheme 2.33 Macrodilactones with several oxygen functionalities in a *1,4*-relationship

Functional group oriented retrosynthesis for the synthesis of hydroxy acid **11** would indicate the formation of skeletal bonds between the functional groups. No matter where the cuts are made, umpoled synthons are required, as can be seen from the nearly 20 syntheses that have been carried out on this class of natural products (Scheme 2.34). In these syntheses the $\mathbf{a}^1 - \mathbf{d}^3$ and $\mathbf{a}^3 - \mathbf{d}^1$ combinations have been preferred over the equally possible $\mathbf{a}^2 - \mathbf{d}^2$ combination [49].

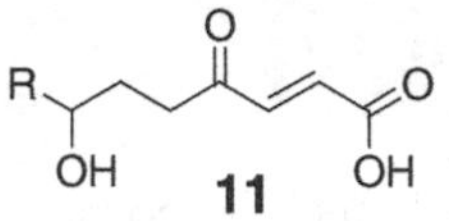

Bond formation between the functional groups requires the application of umpoled synthons in each of the possible combinations $\mathbf{a}^1$-$\mathbf{d}^3$,$\mathbf{a}^2$-$\mathbf{d}^2$,$\mathbf{a}^3$-$\mathbf{d}^1$

Scheme 2.34 Hydroxy acid with *1,4*-relationship of oxyfunctionalities

Scheme 2.35 highlights some $\mathbf{a}^1 - \mathbf{d}^3$ combinations that have been used in the synthesis of **11**. The most efficient method appears to be the Pd(0)-mediated coupling of an acid chloride with a vinylstannane moiety [50, 51].

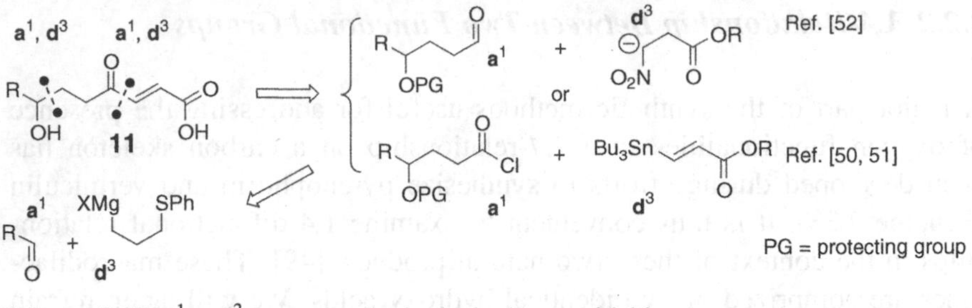

a^1, d^3 a^1, d^3 d^3 Ref. [52]

(Scheme 2.35 structures)

a^1 XMg   SPh

PG = protecting group

Scheme 2.35 $a^1 - d^3$ routes to obtain the hydroxy acid **11**

The synthesis shown in Scheme 2.36 [53] uses a building block **12**, which has umpoled reactivity (d^1 and d^3) on both of its ends. This allows for the indicated $a^3 - d^1-$ and $d^3 - a^1-$bond formations to be accomplished.

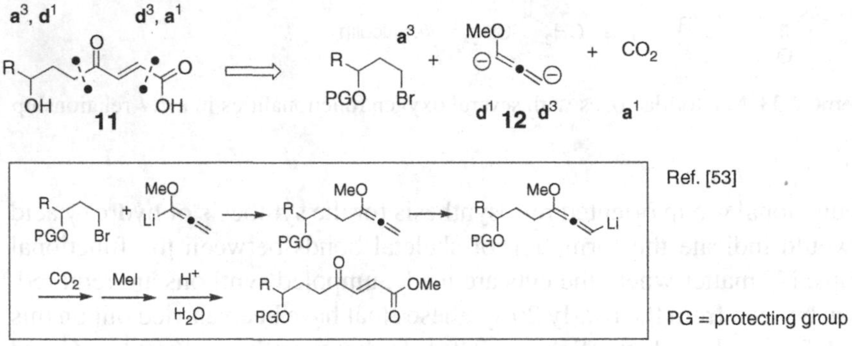

a^3, d^1 d^3, a^1 a^3 MeO $+ CO_2$

(Scheme 2.36 structures)

d^1 **12** d^3 a^1

Ref. [53]

(reaction scheme: MeO ... → ... → ... Li)

CO_2 MeI H^+ / H_2O → (product) OMe

PG = protecting group

Scheme 2.36 $a^3 - d^1$ routes to obtain the hydroxy acid **11**

Regarding the use of a (latent) a^2-synthon the following synthesis [54] (Scheme 2.37) of hydroxy acid **11** is of interest.

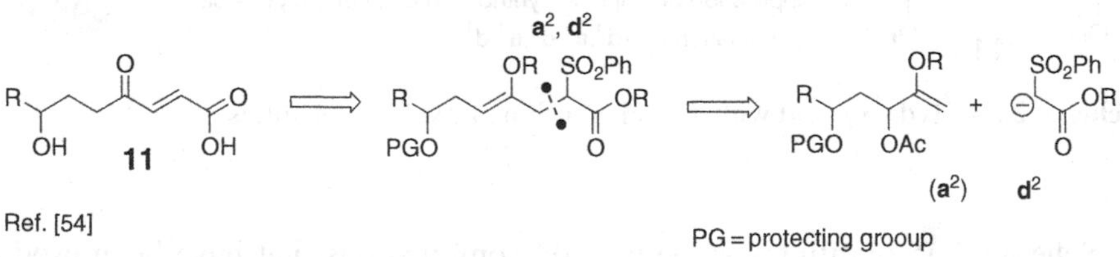

a^2, d^2

OR SO_2Ph OR SO_2Ph

(Scheme 2.37 structures)

(a^2) d^2

Ref. [54]

PG = protecting grooup

Scheme 2.37 Use of an a^2-synthon during a synthesis of the hydroxy acid **11**

One should note that in this and previous examples, the building blocks were chosen with functionality to allow the ready formation of the C-2/C-3 double bond (by elimination of $PhSO_2^-$ or of NO_2^- [52]).

Synthetic approaches in which two skeletal bonds are formed "piece-meal" between a single pair of functional groups are more circumstantial and, hence, less attractive (Scheme 2.38). They instead reflect the change to a building block oriented synthesis strategy.

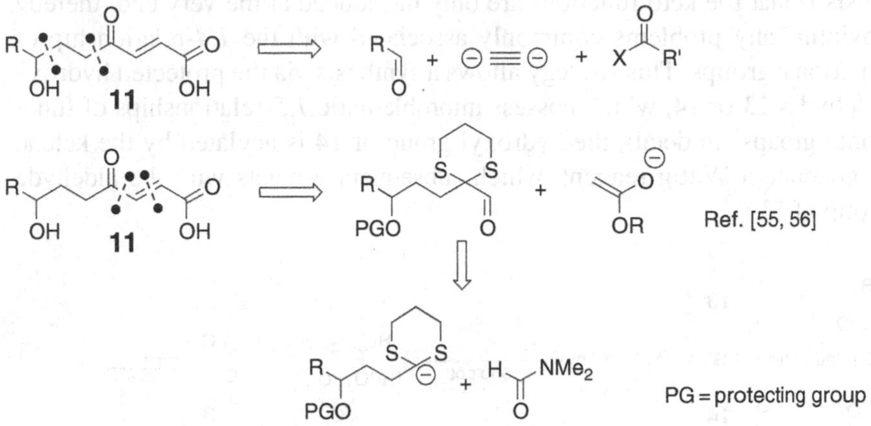

Scheme 2.38 C_2- respectively C_1-piecemeal assembly of the hydroxy acid **11**

Building block based syntheses of hydroxy acid **11** are displayed in Scheme 2.39. Each building block is highlighted. Other 1,4-difunctionalized building blocks like the ones used in the synthesis of **11** can be found in reference [7].

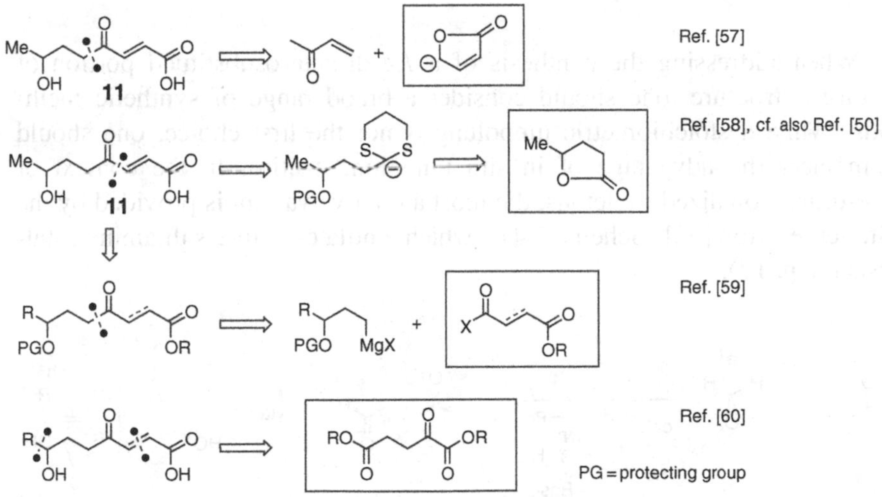

Scheme 2.39 Building block oriented syntheses of hydroxy acid **11**

The syntheses of the hydroxy acid **11** subunit of pyrenophorin presented so far illustrate reliable, orthodox approaches, each of which allows one to reach the target with justifiable effort. Nevertheless, there is always room for more creative approaches. One such approach [61] (directed to an analogue of pyrenphorin) is shown in Scheme 2.40. The surprising feature of this synthesis is that the keto functions are only introduced at the very end, thereby obviating any problems commonly associated with the *1,4*-relationship of functional groups. This strategy allows a synthesis via the protected hydroxy aldehydes **13** or **14**, which possess unproblematic *1,5*-relationships of functional groups. In detail, the hydroxyl group of **14** is acylated by the ketene to generate a Wittig reagent, which subsequently reacts with the aldehyde group of **13**.

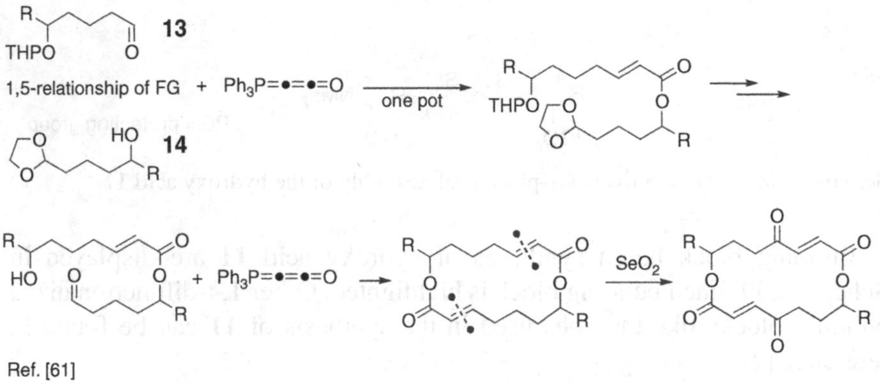

Ref. [61]

Scheme 2.40 Bestmann's synthesis of the pyrenophorin skeleton

When addressing the synthesis of a *1,4*-di-heterosubstitued portion of a target structure, one should consider a braod range of synthetic methods. While a stoichiometric umpolung is not the first choice, one should remember the advantage of in situ umpolung tactics. In the context of *1,4*-difunctionalized skeletons, the most attractive variant is provided by the Stetter reaction [62] (Scheme 2.41), which emulates nature's thiamine catalysis (cf. p. 17).

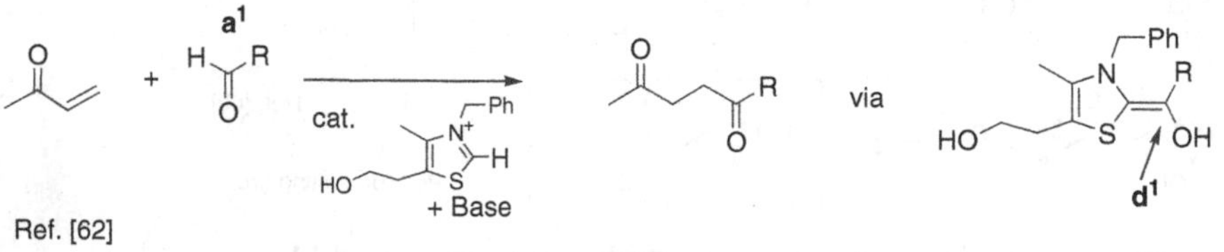

Ref. [62]

Scheme 2.41 Stetter's in situ umpolung of aldehydes

Another in situ umpolung that fits here is the one electron redox umpolung of β-dicarbonyl compounds, mentioned on p. 22 and the related oxidation of enolates [63] or of enamines [64] (Scheme 2.42).

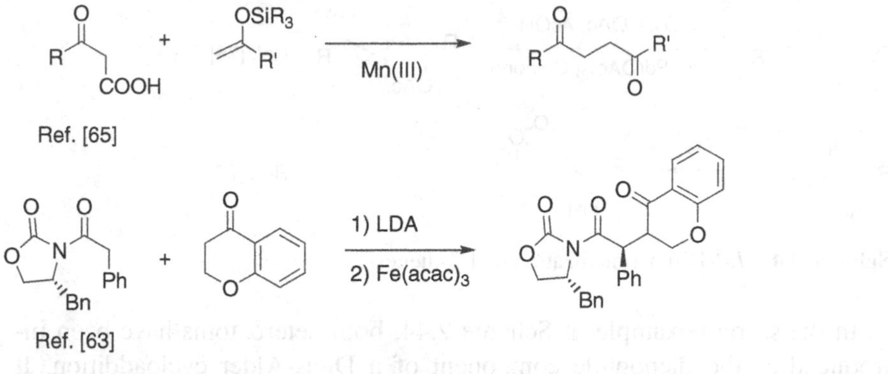

Ref. [65]

Ref. [63]

Scheme 2.42 In situ umpolung of enolates by one-electron oxidation

When one looks further at how structures with a *1,4*-relationship of heteroatoms may be accessed, one should consider either the sequence of a Claisen ester rearrangement followed by Wacker oxidation, or the iodolactonization or direct [66] lactonization reaction of the carbon-carbon double bond (Scheme 2.43). For an application in the context of pyrenophorin synthesis, see reference [67]. Note that in the course of this sequence the heteroatom, which was at C-1 of the allylic alcohol, is moved over to C-2.

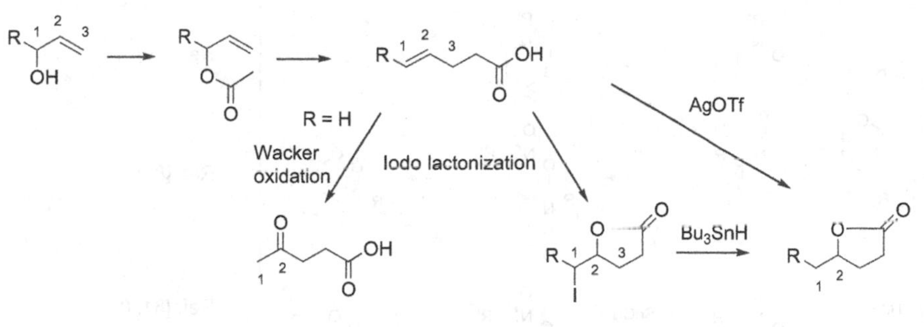

Scheme 2.43 Skeleton bond-forming reaction sequence based on a Claisen rearrangement to generate *1,4*-difunctionality

In Bestmann's synthesis of a pyrenophorin analogue (Scheme 2.40; cf. p. 30), functionality was introduced by an oxidation reaction after assembling the molecular skeleton. In principle, it is possible and also advantageous to attach a heteroatom at each end of a 1,3-diene in order to generate a *1,4*-difunctionality. Methods that serve this purpose are the Pd(II)-mediated oxyfunctionalization according to Bäckval [68] or the addition of singlet oxygen in a hetero-Diels-Alder cycloaddition [69]. Further heteroatom dienophiles of interest are ROOC-N=N-COOR, O=N-COOR [70, 71], O=N-Ph [72, 73], and RN=S=O [74, 75] (Scheme 2.44).

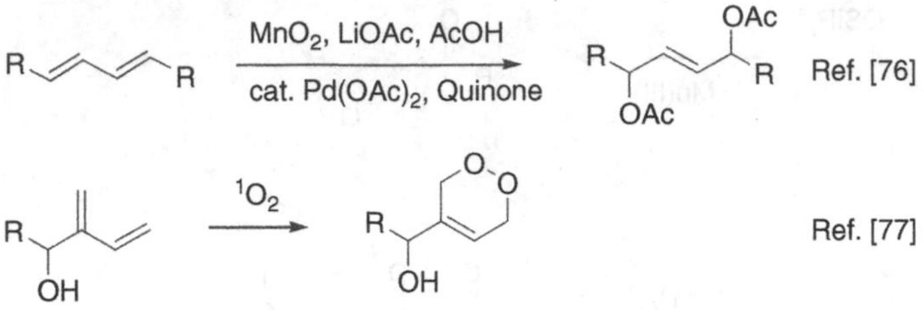

Scheme 2.44 *1,4*-Difunctionalization of 1,3-dienes

In the second example of Scheme 2.44, both heteroatoms have been introduced as the dienophile component of a Diels-Alder cycloaddition. It is likewise possible to use heteroatom-based dienes to introduce a 1,4-difunctionality via a Diels-Alder cycloaddition (Scheme 2.45).

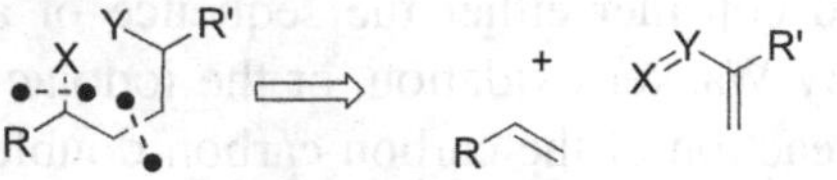

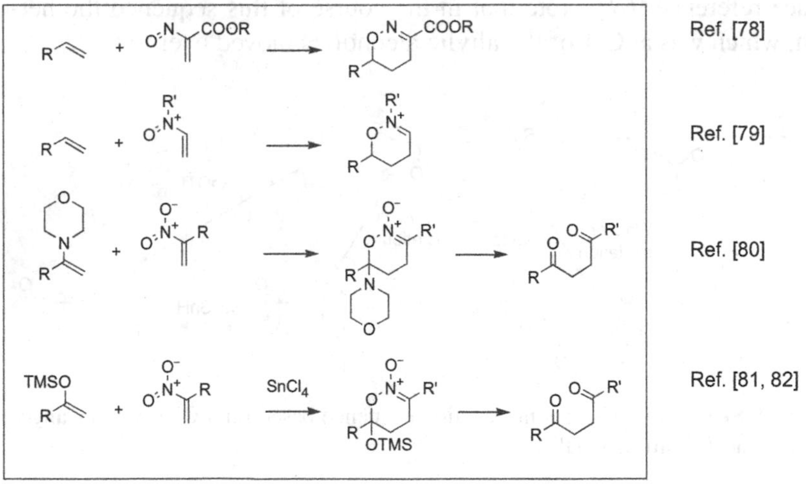

Scheme 2.45 Installing both heteroatoms of a *1,4*-difunctionality by a Diels-Alder cycloaddition using a dihetero-1,3-diene

A conceptually related approach is to add ketene or dichloroketene to an alkene, followed by oxidative ring enlargement of the resulting cyclobutanone (Scheme 2.46).

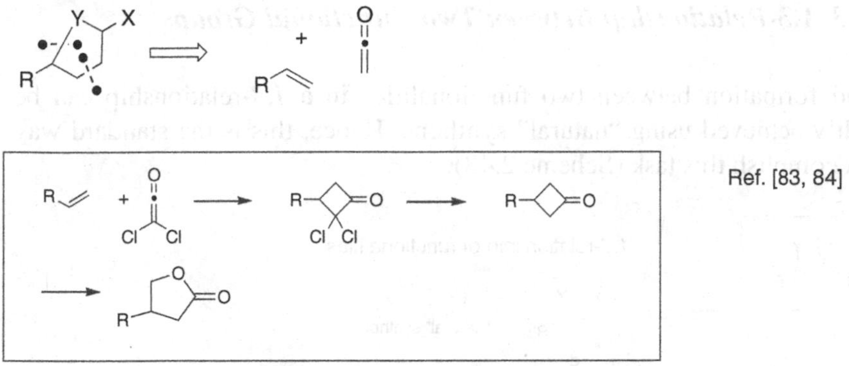

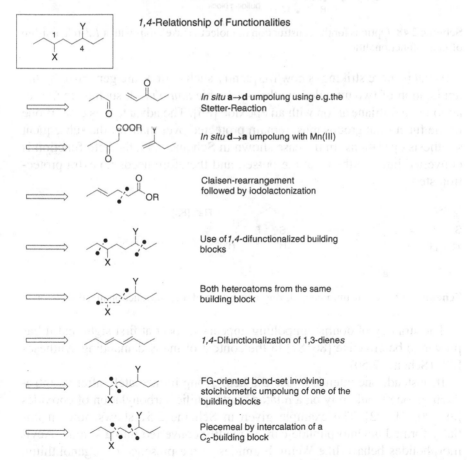

Scheme 2.46 *1,4*-Difunctionalized skeletons via cyclobutanones

The various options to attain molecular skeletons with a *1,4*-relationship of functionalities are summarized and ranked (top = best) in Scheme 2.47:

Scheme 2.47 Options for the construction of molecular skeletons with a *1,4*-relationship of heterofunctionalities

2.2.3 1,3-Relationship Between Two Functional Groups

Bond formation between two functionalities in a *1,3*-relationship can be readily achieved using "natural" synthons. Hence, this is the standard way to accomplish this task (Scheme 2.48).

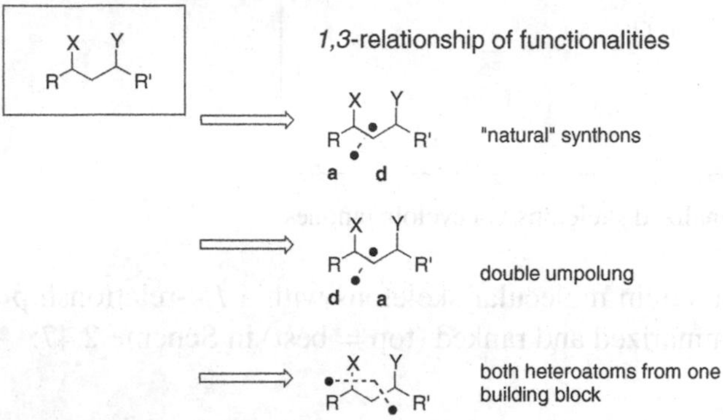

Scheme 2.48 Options for the construction of molecular skeletons with a *1,3*-relationship of heterofunctionalities

What is more striking is how frequently such entities are generated by the application of two umpoled synthons (*double umpolung*), such as in the reaction of a dithiane anion with an epoxide [85]. The advantage is clear if one of the functional groups must remain protected over many of the subsequent synthesis operations. In the case shown in Scheme 2.49, the keto function is converted into a dithiane at the outset, and therefore needs no extra protection step.

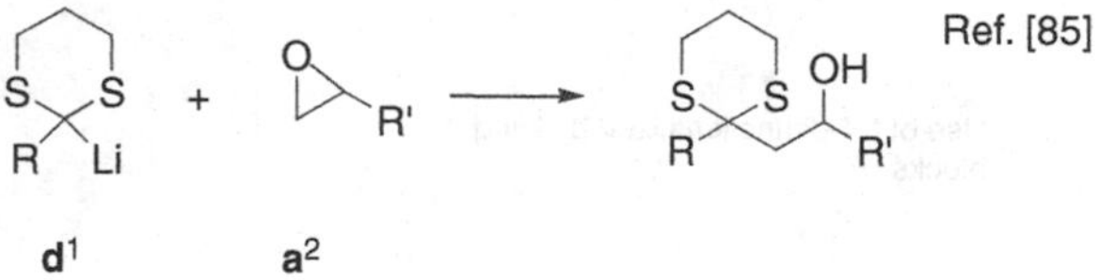

Scheme 2.49 Double umpolung during generation of a *1,3*-difunctionalized skeleton

The strategy of double umpolung appears suspect at first sight, but it has proven to be effective [86, 87] in the context of many demanding syntheses [88] (Scheme 2.50).

If, instead, one intends to use the keto group immediately after its introduction, one would rely on a related nucleophilic carbonylation of epoxides [89, 90, 91, 92]. The example given in Scheme 2.51 shows how an initially formed acylmorpholide can be readily converted into a ketone. Acylmorpholides behave like Weinreb amides in the presence of organolithium reagents, making ketones (in this case a β-hydroxy-ketone) available [93].

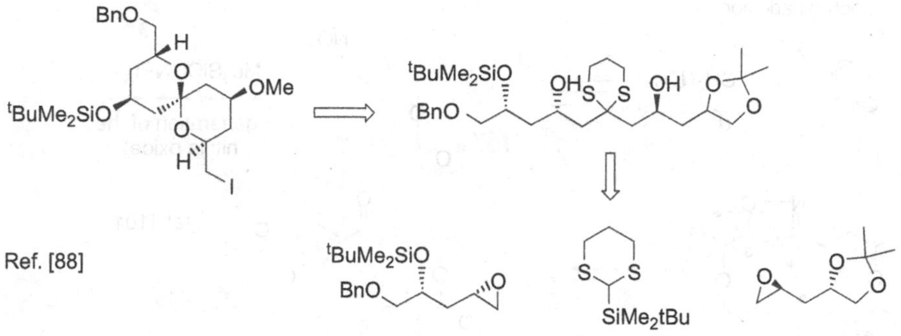

Ref. [88]

Scheme 2.50 Double umpolung in a challenging context

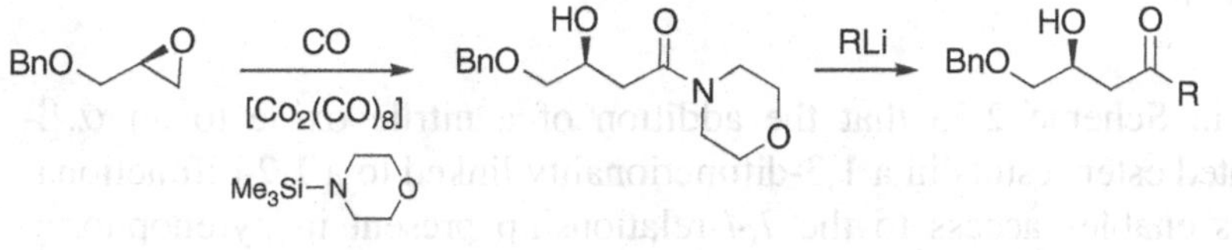

Ref. [89]

Scheme 2.51 Double umpolung via nucleophilic carbonylation

Aside from the standard use of natural synthons and the routes via double umpolung, there is the possibility of enlisting 1,3-dipolar cycloadditions to form *1,3*-difunctionalized molecular skeletons. Typical examples include the addition of nitrones, silyl nitronates [94], or nitrile oxides to alkenes. The initial products of the latter cycloaddition are isoxazolines, which may be refunctionalized in various ways (Scheme 2.52). When this generates sensitive functionalities, refunctionalization may be postponed until later in the synthesis sequence [95].

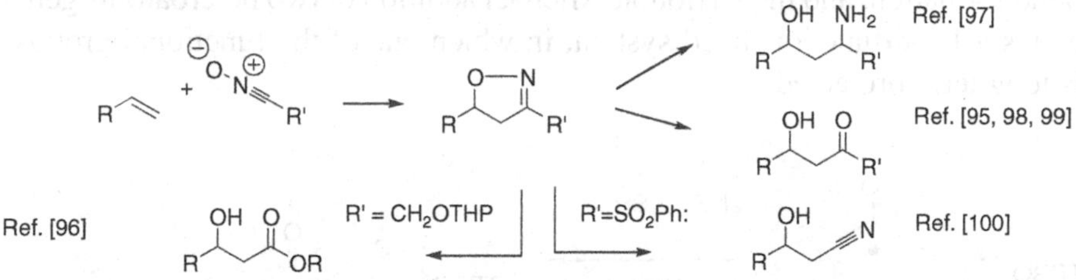

Ref. [96]   R' = CH$_2$OTHP   R'=SO$_2$Ph:

Ref. [97]
Ref. [95, 98, 99]
Ref. [100]

Scheme 2.52 1,3-Dipolar cycloaddition of nitrile oxides furnishing *1,3*-difunctionalized molecular skeletons

The versatile nitrile oxide cycloaddition has been exploited in another highly efficient synthesis of pyrenophorin [101] (Scheme 2.53).

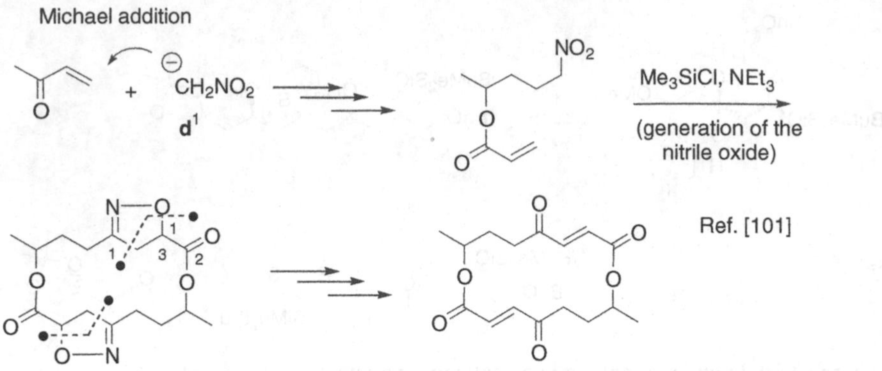

Michael addition

Scheme 2.53 1,3-Dipolar cycloaddition of nitrile oxides as key step in a synthesis of pyrenophorin

Note in Scheme 2.53 that the addition of a nitrile oxide to an α, β-unsaturated ester results in a 1,3-difunctionality linked to a 1,2-difunctionality. This enables access to the *1,4*-relationship present in pyrenophorin. A similar outcome is obtained with the Kanemasa variant [102] of the cycloaddition of nitrile oxide with the double bond of allylic alcohols. This variant is impressive because of its high regio- and stereo-selectivity [95, 103, 104] (Scheme 2.54).

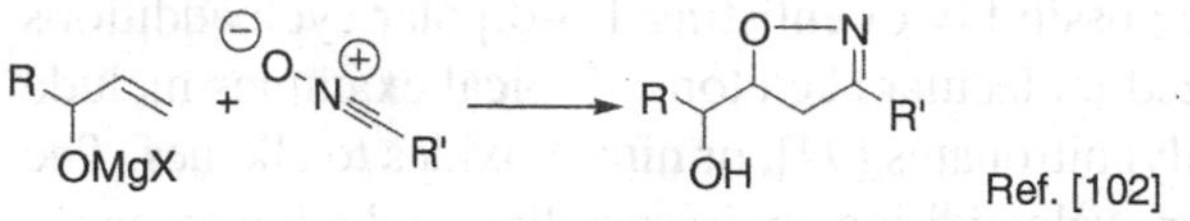

Scheme 2.54 Kanemasa variant of the addition of nitrile oxides to allylic alcohols

In terms of bond sets, the following sequence (Scheme 2.55) is equivalent to that of 1,3-dipolar cycloadditions. First an ynone is created by skeletal bond formation, and then a double Michael addition of two heteroatoms generates a 1,3-difunctionalized system, in which one of the functional groups is long-term protected.

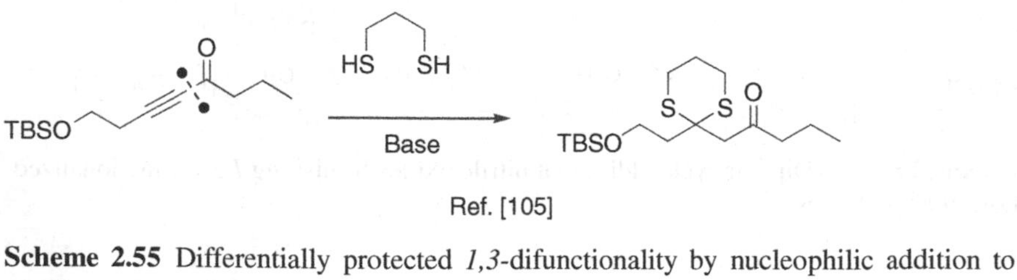

Scheme 2.55 Differentially protected *1,3*-difunctionality by nucleophilic addition to ynones

2.2.4 1,5-(and 1,6-) Relationship Between Two Functional Groups

For the generation of molecular skeletons with a *1,5*-relationship of functionalities, the approach combining the natural $\mathbf{d}^2$-synthon with a natural $\mathbf{a}^3$-synthon is the predominant tactic. The alternative, introduction of both heteroatoms via a single building block, e.g., in a hetero-Diels-Alder cycloaddition [106] (Scheme 2.56), is not yet fully developed and frequently suffers from competing polymerization of the enone partner.

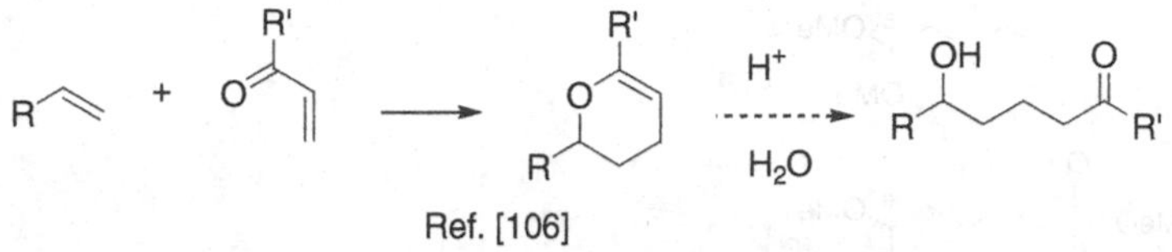

Ref. [106]

Scheme 2.56 Hetero-Diels-Alder cycloaddition en route to *1,5*-difunctionalized skeletons

Another prominent route to develop 1,5-difunctionlized molecular skeletons is the oxidative cleavage of cyclopentene derivatives (Scheme 2.57). In Corey's terminology of retrosynthesis this is the "reconnect" operation (reconnection of the two functionalities). Other reactions that fall under this heading include the Baeyer-Villiger oxidation and the Beckmann rearrangement of cyclopentanones.

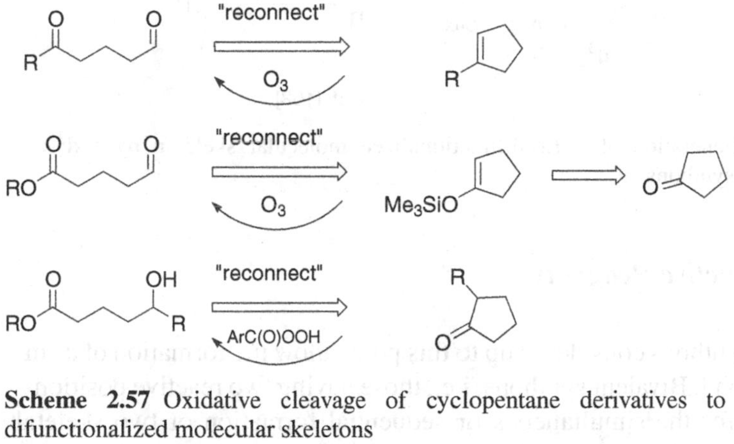

Scheme 2.57 Oxidative cleavage of cyclopentane derivatives to produce *1,5*-difunctionalized molecular skeletons

The transformations shown in Scheme 2.57 are refunctionalization reactions. They are worthy of consideration when suitable cyclopentene or cyclopentanone building blocks are readily available.

Correspondingly, 1,6-difunctionalized molecular skeletons can be derived by oxidative cleavage of cyclohexene derivatives. This is a widely used method, as it allows for differentiation of the functionalities at the 1- and 6-positions. Detailed methods for doing this can be found in reference [107] (Scheme 2.58).

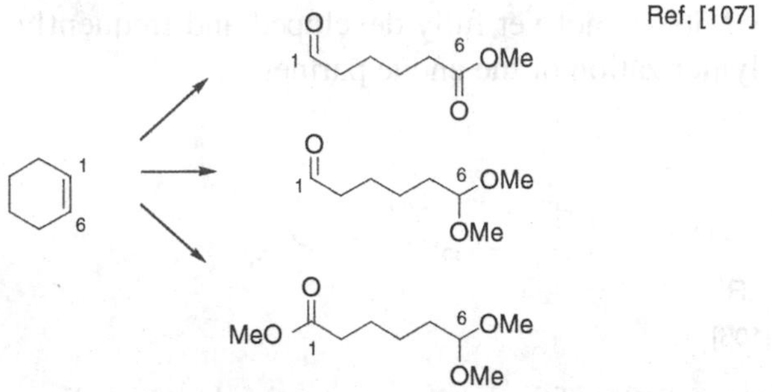

Scheme 2.58 Oxidative cleavage of cyclohexene (derivatives) to generate end group differentiated *1,6*-difunctionalized molecular skeletons

In the event that one desires to generate a *1,6*-difuctionalized molecular skeleton via bond formation between functionalities, there remains the single option: use a **d^3-a^3**-combination of synthons (Scheme 2.59).

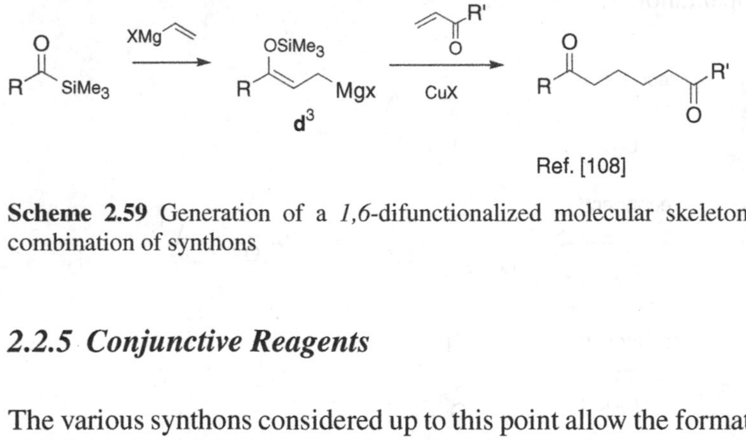

Scheme 2.59 Generation of a *1,6*-difunctionalized molecular skeleton by a **d^3-a^3**-combination of synthons

2.2.5 Conjunctive Reagents

The various synthons considered up to this point allow the formation of a single skeletal bond. Bivalent synthons, i.e., those having two reactive positions, would allow for the simultaneous or sequential formation of two skeletal bonds. An example can be found in Scheme 2.38 (see p. 29), where an acetylene has been placed as a C_2-unit between two other synthons. Such bivalent synthons have been referred to by Trost [109] as *conjunctive reagents*, and by Seebach [110] as *multicoupling reagents*. The simplest conjunctive reagent is represented by a CH_2-unit, which ought to be available in three modifications of different polarities to allow universal application (Scheme 2.60).

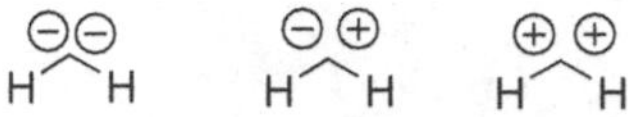

$$\overset{\ominus\ominus}{H\diagup H} \qquad \overset{\ominus\ \oplus}{H\diagup H} \qquad \overset{\oplus\ \oplus}{H\diagup H}$$

Scheme 2.60 Bivalent methylene synthons as conjunctive reagents

A dianionic methylene conjunctive reagent requires groups Y and Z, which can stabilize the adjacent negative charge [111, 112]. Sulfonyl-stabilized carbanions of this type can be alkylated twice. Afterwards, the sulfonyl groups must be removed in a subsequent reduction step in order to reveal the intended CH_2-unit (Scheme 2.61).

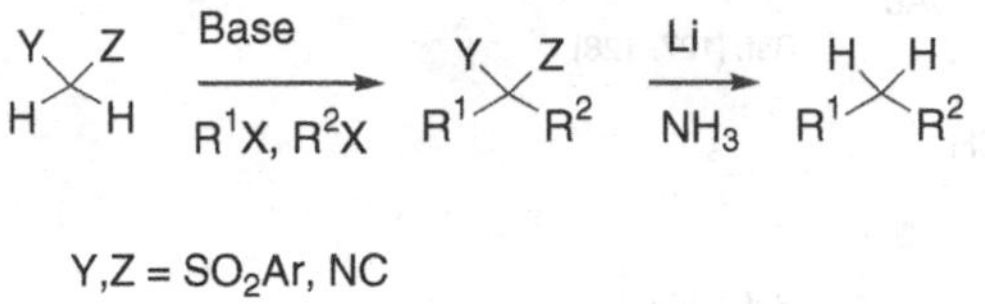

$$\underset{H\diagup H}{\overset{Y\diagdown Z}{\diagdown}} \ \xrightarrow[R^1X,\ R^2X]{Base}\ \underset{R^1\diagup R^2}{\overset{Y\diagdown Z}{\diagdown}}\ \xrightarrow[NH_3]{Li}\ \underset{R^1\diagup R^2}{\overset{H\diagdown H}{\diagdown}}$$

Y,Z = SO_2Ar, NC

Scheme 2.61 *1,1*-Bivalent methylene as the simplest dianionic conjunctive reagent

Inserting just a methylene unit is not particularly exciting from a synthetic point of view. It is much more interesting to consider functionalized conjunctive reagents. An important synthon of this sort is the "carbonyl-dianion." Of the reagents corresponding to this synthon, dithiane is used most frequently [113] (Scheme 2.62).

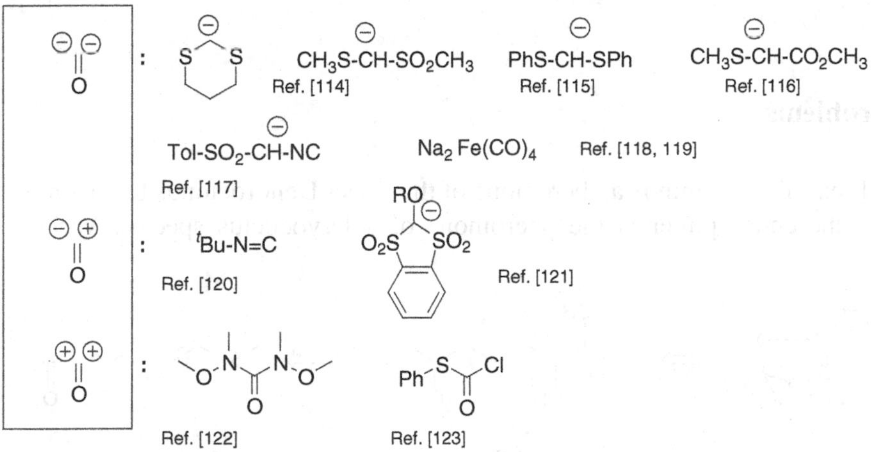

Scheme 2.62 *1,1*-Bivalent carbonyl synthons and corresponding reagents

Other useful conjunctive reagents are listed in Scheme 2.63 (cf. also Scheme 5.7 on p. 78).

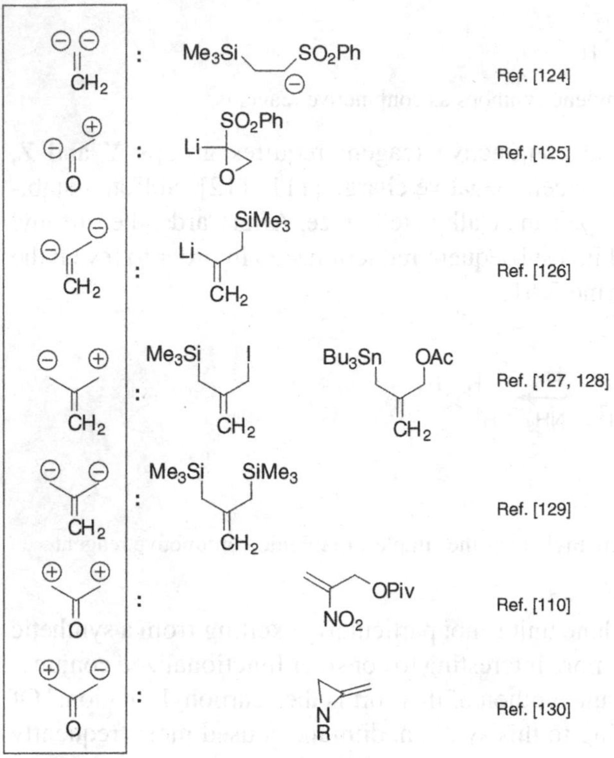

Scheme 2.63 Examples of *1,1-*, *1,2-*, and *1,3*-bivalent conjunctive reagents

Problems

2.1 exo-Brevicomin is a pheromone of the insect Dendroctonus brevicomis; the endo-epimer is the pheromone of a Dryocoetus species. Various

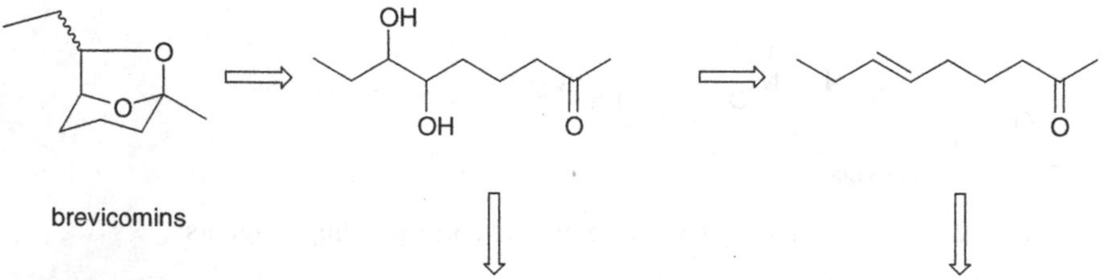

brevicomins

Scheme 2.64 Retrosynthesis of the brevicomins

routes to the syntheses of the brevicomins have been explored [131]. Most syntheses proceed via the dihydroxyketone shown in Scheme 2.64.

Develop a retrosynthesis of the brevicomins along these lines and discuss the pros and cons of going retrosynthetically back to a double bond as the profunctionality of the diol unit (consult references [132, 133, 134, 135, 136]).

2.2 A versatile intermediate for the synthesis of indolizidine alkaloids is the compound shown in Scheme 2.65.

Scheme 2.65 Intermediate for indolizidine alkaloid synthesis

Follow the skeletal bonds around the molecule. Which distances between heteroatom groups can you delineate? Which of them are unproblematic in synthesis? Consider the carbon-carbon bonds marked A–C for construction of this molecule. Evaluate the polarity options for making these bonds.

References

1. E. J. Corey, X.-M. Cheng *The Logic of Chemical Synthesis*, J. Wiley & Sons, New York, **1989**.
2. D. Seebach, *Angew. Chem. Int. Ed. Engl.* **1979**, *18*, 239–258. (*Angew. Chem.* **1979**, *91*, 259–278).
3. R. C. Larock *Comprehensive Organic Transformations*, Wiley-VCH, New York, **1989**.
4. D. A. Evans, G. C. Andrews, *Acc. Chem. Res.* **1974**, *7*, 147–155
5. F. Serratosa *Organic Chemistry in Action*, 1st ed., Elsevier, Amsterdam, **1990**
6. G. S. Zweifel, M. H. Nantz *Modern Organic Synthesis: An Introduction*, W. H. Freeman and Co., New York, **2007**.
7. S. Warren *Organic Synthesis: The Disconnection Approach*, J. Wiley & Sons, Chichester, **1982**.
8. E. J. Corey, *Pure Appl. Chem.* **1967**, *14*, 19–37.
9. W. A. Smit, A. F. Bochkov, R. Caple *Organic Synthesis: The Science behind the Art*, Royal Society of Chemistry, **1998**, Chapt. 2.16, p. 152.
10. D. Seebach *Organolithium Compounds in Organic Synthesis* in *New Applications of Organometallic Reagents in Organic Synthesis* (Ed.: D. Seyferth), Elsevier, Amsterdam, **1976**, pp. 1–92.

11. M. Braun *Houben Weyl, Methoden der Organischen Chemie* vol. E19d, (Ed.: M. Hanack), G. Thieme, Stuttgart, **1993**, pp. 853–1138.
12. R. W. Stevens, T. Mukaiyama, *Chem. Lett.* **1985**, 851–854.
13. R. Fernández, J. M. Lassaletta, *Synlett.* **2000**, 1228–1240.
14. M. Yus, J. Ortiz, C. Najera, *ARKIVOC* **2002**(v), 38–47.
15. W. C. Still, *J. Am. Chem. Soc.* **1978**, *100*, 1481–1487.
16. W. C. Still, C. Sreekumar, *J. Am. Chem. Soc.* **1980**, *102*, 1201–1202.
17. J. M. Chong, E. K. Mar, *Tetrahedron Lett.* **1990**, *31*, 1981–1984.
18. J. C. Stowell, *Chem. Rev.* **1984**, *84*, 409–435.
19. D. Hoppe, *Angew. Chem., Int. Ed. Engl.* **1984**, *23*, 932–948. (*Angew. Chem.* **1984**, *96*, 930–946).
20. H. Nakahira, M. Ikebe, Y. Oku, N. Sonoda, T. Fukuyama, I. Ryu, *Tetrahedron* **2005**, *61*, 3383–3392.
21. I. Ryu, G. Yamamura, S. Omura, S. Minakata, M. Komatsu, *Tetrahedron Lett.* **2006**, *47*, 2283–2286.
22. J. S. Johnson, *Angew. Chem., Int. Ed.* **2004**, *43*, 1326–1328. (*Angew. Chem.* **2004**, *116*, 1348–1350).
23. B. K. Banik, *Eur. J. Org. Chem.* **2002**, 2431–2444.
24. H. C. Aspinall, N. Greeves, C. Valla, *Org. Lett.* **2005**, *7*, 1919–1922.
25. T. Wirth, *Angew. Chem., Int. Ed. Engl.* **1996**, *35*, 61–63. (*Angew. Chem.* **1996**, *108*, 65–67).
26. Y. Taniguchi, M. Nakahashi, T. Kuno, M. Tsuno, Y. Makioka, K. Takaki, Y. Fujiwara, *Tetrahedron Lett.* **1994**, *35*, 4111–4114.
27. D. S. Surry, D. R. Spring, *Chem. Soc. Rev.* **2006**, *35*, 218–225.
28. T. Shono, N. Kise, T. Fujimoto, A. Yamanami, R. Nomura, *J. Org. Chem.* **1994**, *59*, 1730–1740.
29. R. K. Dieter, S.-J. Li, N. Chen, *J. Org. Chem.* **2004**, *69*, 2867–2870.
30. Y.-I. Yoshida, M. Itoh, S. Isoe, *J. Chem. Soc., Chem. Commun.* **1993**, 547–549.
31. Y.-I. Yoshida, Y. Ishichi, S. Isoe, *J. Am. Chem. Soc.* **1992**, *114*, 7594–7595.
32. J. B. Sperry, D. L. Wright, *Chem. Soc. Rev.* **2006**, *35*, 605–621.
33. K. Naraska, Y. Kohno, S. Shimada, *Chem. Lett.* **1993**, *22*, 125–128.
34. K. Otsubo, J. Inanaga, M. Yamaguchi, *Tetrahedron Lett.* **1986**, *27*, 5763–5764.
35. E. J. Enholm, H. Satici, A. Trivellas, *J. Org. Chem.* **1989**, *54*, 5841–5843.
36. G. Masson, P. Cividino, S. Py, Y. Vallée, *Angew. Chem., Int. Ed.* **2003**, *42*, 2265–2268. (*Angew. Chem.* **2003**, *115*, 2367–2370).
37. E. J. Corey, A. K. Ghosh, *Chem. Lett.* **1987**, *16*, 223–226.
38. B. B. Snider, *Chem. Rev.* **1996**, *96*, 339–363.
39. T. Linker, *J. Prakt. Chem.* **1997**, *339*, 488–492.
40. F. A. Davis, B.-C. Chen, *Chem. Rev.* **1992**, *92*, 919–934.
41. C. Greck, J. P. Gênet, *Synlett* **1997**, 741–748.
42. P. Dembech, G. Seconi, A. Ricci, *Chem. Eur. J.* **2000**, *6*, 1281–1286.
43. C. D. Johnson, *Acc. Chem. Res.* **1993**, *26*, 476–482.
44. P. C. Wälchli, C. H. Eugster, *Helv. Chim. Acta* **1978**, *61*, 885–898.
45. S. R. Angle, R. M. Henry, *J. Org. Chem.* **1998**, *63*, 7490–7497.
46. Y. Hirai, J. Watanabe, T. Nozaki, H. Yokoyama, S. Yamaguchi, *J. Org. Chem.* **1997**, *62*, 776–777.
47. O. Muraoka, B.-Z. Zheng, K. Okumura, G. Tanabe, T. Momose, C. H. Eugster, *J. Chem. Soc., Perkin Trans. 1,* **1996**, 1567–1575.

48. B. Nader, T. R. Bailey, R. W. Franck, S. M. Weinreb, *J. Am. Chem. Soc.* **1981**, *103*, 7573–7580.
49. R. S. Mali, M. Pohmakotr, B. Weidmann, D. Seebach, *Liebigs Ann. Chem.* **1981**, 2272–2284.
50. J. W. Labadie, D. Tueting, J. K. Stille, *J. Org. Chem.* **1983**, *48*, 4634–4642.
51. J. E. Baldwin, R. M. Adlington, S. H. Ramcharitar, *Synlett* **1992**, 875–877.
52. P. Bakuzis, M. L. F. Bakuzis, T. F. Weingartner, *Tetrahedron Lett.* **1978**, *19*, 2371–2374.
53. F. Derguini, G. Linstrumelle, *Tetrahedron Lett.* **1984**, *25*, 5763–5766.
54. B. M. Trost, F. W. Gowland, *J. Org. Chem.* **1979**, *44*, 3448–3450.
55. B. Seuring, D. Seebach, *Liebigs Ann. Chem.* **1978**, 2044–2073.
56. K. F. Burri, R. A. Cardone, W. Y. Chen, P. Rosen, *J. Am. Chem. Soc.* **1978**, *100*, 7069–7071.
57. M. Asaoka, N. Yanagida, N. Sugimura, T. Takei, *Bull. Chem. Soc. Jpn.* **1980**, *53*, 1061–1064.
58. E. W. Colvin, T. A. Purcell, R. A. Raphael, *J. Chem. Soc., Perkin Trans. 1,* **1976**, 1718–1722.
59. H. Gerlach, K. Oertle, A. Thalmann, *Helv. Chim. Acta* **1977**, *60*, 2860–2865.
60. E. J. Corey, K. C. Nicolaou, T. Toru, *J. Am. Chem. Soc.* **1975**, *97*, 2287–2288.
61. H. J. Bestmann, R. Schobert, *Angew. Chem., Int. Ed. Engl.* **1985**, *24*, 791–792. (*Angew. Chem.* **1985**, *97*, 784–785).
62. H. Stetter, *Angew. Chem., Int. Ed. Engl.* **1976**, *15*, 639–647. (*Angew. Chem.* **1976**, *88*, 695–704).
63. P. S. Baran, M. P. DeMartino, *Angew. Chem., Int. Ed.* **2006**, *45*, 7083–7086. (*Angew. Chem.* **2006**, *118*, 7241–7244).
64. H.-Y. Jang, J.-B. Hong, D. W. C. MacMillan, *J. Am. Chem. Soc.* **2007**, *129*, 7004–7005.
65. K. Narasaka, N. Miyoshi, K. Iwakura, T. Okauchi, *Chem. Lett.* **1989**, *18*, 2169–2172.
66. C.-G. Yang, N. W. Reich, Z. Shi, C. He, *Org. Lett.* **2005**, *7*, 4553–4556.
67. T. A. Hase, A. Ourila, C. Holmberg, *J. Org. Chem.* **1981**, *46*, 3137–3139.
68. J.-E. Bäckval, *Bull. Soc. Chim. Fr.* **1987**, 665–670.
69. H. H. Wasserman, J. L. Ives, *Tetrahedron* **1981**, *37*, 1825–1852.
70. A. Defoin, H. Fritz, G. Geffroy, G. Streith, *Tetrahedron Lett.* **1986**, *27*, 4727–4730.
71. D. L. Boger, M. Patel, F. Takusagawa, *J. Org. Chem.* **1985**, *50*, 1911–1916.
72. D. J. Dixon, S. V. Ley, D. J. Reynolds, *Angew. Chem., Int. Ed.* **2000**, *39*, 3622–3626. (*Angew. Chem.* **2000**, *112*, 3768–3772).
73. Y. Yamamoto, H. Yamamoto, *Angew. Chem., Int. Ed.* **2005**, *44*, 7082–7085. (*Angew. Chem.* **2005**, *117*, 7244–7247).
74. R. S. Garigipati, A. J. Freyer, R. R. Whittle, S. M. Weinreb, *J. Am. Chem. Soc.* **1984**, *106*, 7861–7867.
75. J. K. Whitesell, D. James, J. F. Carpenter, *J. Chem. Soc., Chem. Commun.* **1985**, 1449–1450.
76. J.-E. Baeckvall, S. E. Byström, R. E. Nordberg, *J. Org. Chem.* **1984**, *49*, 4619–4631.
77. K. Kondo, M. Matsumoto, *Tetrahedron Lett.* **1976**, *17*, 4363–4366.
78. T. L. Gilchrist, D. A. Lingham, T. G. Roberts, *J. Chem. Soc., Chem. Commun.* **1979**, 1089–1090.

79. U. M. Kempe, T. K. Das Gupta, K. Blatt, P. Gygax, D. Felix, A. Eschenmoser, *Helv. Chim. Acta* **1972**, *55*, 2187–2198.
80. F. Felluga, P. Nitti, G. Pitacco, E. Valentin, *Tetrahedron* **1989**, *45*, 2099–2108.
81. M. Miyashita, T. Yanami, A. Yoshikoshi, *J. Am. Chem. Soc.* **1976**, *98*, 4679–4681.
82. S. E. Denmark, A. Thorarensen, *Chem. Rev.* **1996**, *96*, 137–165.
83. J. A. Hyatt, P. W. Raynolds, *Org. Reactions* **1994**, *45*, 159–646.
84. G. R. Krow, *Org. Reactions* **1993**, *43*, 251–798.
85. P. C. Bulman Page, M. B. van Niel, J. C. Prodger, *Tetrahedron* **1989**, *45*, 7643–7677.
86. M. Yus, C. Nájera, F. Foubelo, *Tetrahedron* **2003**, *59*, 6147–6212.
87. A. B. Smith III, S. M. Pitram, A. M. Boldi, M. J. Gaunt, C. Sfouggatakis, W. H. Moser, *J. Am. Chem. Soc.* **2003**, *125*, 14435–14445.
88. A. B. Smith III, V. A. Doughty, Q. Lin, L. Zhuang, M. D. McBriar, A. M. Boldi, W. H. Moser, N. Murase, K. Nakayama, M. Sobukawa, *Angew. Chem., Int. Ed.* **2001**, *40*, 191–195. (*Angew. Chem.* **2001**, *113*, 197–201).
89. S. N. Goodman, E. N. Jacobsen, *Angew. Chem., Int. Ed.* **2002**, *41*, 4703–4705. (*Angew. Chem.* **2002**, *114*, 4897–4899).
90. J. A. R. Schmidt, E. B. Lobkovsky, G. W. Coates, *J. Am. Chem. Soc.* **2005**, *127*, 11426–11435.
91. J. W. Kramer, E. B. Lobkovsky, G. W. Coates, *Org. Lett.* **2006**, *8*, 3709–3712.
92. S. E. Denmark, M. Ahmad, *J. Org. Chem.* **2007**, *72*, 9630–9634.
93. M. M. Jackson, C. Leverett, J. F. Toczko, J. C. Roberts, *J. Org. Chem.* **2002**, *67*, 5032–5035.
94. T. Ishikawa, Y. Shimizu, T. Kudoh, S. Saito, *Org. Lett.* **2003**, *5*, 3879–3882.
95. J. W. Bode, E. M. Carreira, *J. Org. Chem.* **2001**, *66*, 6410–6424.
96. A. P. Kozikowski, M. Adamczyk, *J. Org. Chem.* **1983**, *48*, 366–372.
97. V. Jäger, W. Schwab, V. Buss, *Angew. Chem., Int. Ed. Engl.* **1981**, *20*, 601–603. (*Angew. Chem.* **1981**, *93*, 576–578).
98. S. F. Martin, B. Dupre, *Tetrahedron Lett.* **1983**, *24*, 1337–1340.
99. Z. Hong, L. Liu, C.-C. Hsu, C.-H. Wong, *Angew. Chem., Int. Ed.* **2006**, *45*, 7417–7421. (*Angew. Chem.* **2006**, *118*, 7577–7581).
100. P. A. Wade, H. R. Hinney, *J. Am. Chem. Soc.* **1979**, *101*, 1319–1320.
101. M. Asaoka, T. Mukuta, H. Takei, *Tetrahedron Lett.* **1981**, *22*, 735–738.
102. S. Kanemasa, M. Nishiuchi, A. Kamimura, K. Hori, *J. Am. Chem. Soc.* **1994**, *116*, 2324–2339.
103. J. W. Bode, N. Fraefel, D. Muri, E. M. Carreira, *Angew. Chem., Int. Ed.* **2001**, *40*, 2082–2085. (*Angew. Chem.* **2001**, *113*, 2128–2131).
104. A. Fürstner, M. D. B. Fenster, B. Fasching, C. Godbout, K. Radkowski, *Angew. Chem., Int. Ed.* **2006**, *45*, 5510–5515. (*Angew. Chem.* **2006**, *118*, 5636–5641).
105. M. J. Gaunt, H. F. Sneddon, P. R. Hewitt, P. Orsini, D. F. Hook, S. V. Ley, *Org. Biomol. Chem* **2003**, *1*, 15–16.
106. L. F. Tietze, G. Kettschau, *Topics Curr. Chem.* **1997**, *189*, 1–120.
107. S. L. Schreiber, R. E. Claus, J. Reagan, *Tetrahedron Lett.* **1982**, *23*, 3867–3870.
108. J. Enda, T. Matsutani, I. Kuwajima, *Tetrahedron Lett.* **1984**, *25*, 5307–5310.
109. B. M. Trost, D. M. T. Chan, *J. Am. Chem. Soc.* **1979**, *101*, 6429–6432.
110. D. Seebach, P. Knochel, *Helv. Chim. Acta* **1984**, *67*, 261–283.
111. J. Yu, H.-S. Cho, J. R. Falck, *J. Org. Chem.* **1993**, *58*, 5892–5894.
112. E. P. Kündig, A. F. Cunningham jr., *Tetrahedron* **1988**, *44*, 6855–6860.
113. A. B. Smith III, C. M. Adams, *Acc. Chem. Res.* **2004**, *37*, 365–377.

114. K. Ogura, M. Suzuki, J.-I. Watanabe, M. Yamashita, H. Iida, G. I. Tsuchihashi, *Chem. Lett.* **1982**, *11*, 813–814.

115. P. Blatcher, J. I. Grayson, S. Warren, *J. Chem. Soc., Chem. Commun.* **1976**, 547–549.

116. B. M. Trost, Y. Tamaru, *J. Am. Chem. Soc.* **1977**, *99*, 3101–3113.

117. O. Possel, A. M. van Leusen, *Tetrahedron Lett.* **1977**, *18*, 4229–4232.

118. J. P. Collman, *Acc. Chem. Res.* **1975**, *8*, 342–347.

119. J. E. McMurry, A. Andrus, *Tetrahedron Lett.* **1980**, *21*, 4687–4690.

120. G. E. Niznik, W. H. Morrison III, H. M. Walborsky, *J. Org. Chem.* **1974**, *39*, 600–604.

121. B. M. Trost, P. Quayle, *J. Am. Chem. Soc.* **1984**, *106*, 2469–2471.

122. W. L. Whipple, H. J. Reich, *J. Org. Chem.* **1991**, *56*, 2911–2912.

123. C. Cardellicchio, V. Fiandanese, G. Marchese, L. Ronzini, *Tetrahedron Lett.* **1985**, *26*, 3595–3598.

124. T. P. Meagher, L. Yet, C.-N. Hsiao, H. Shechter, *J. Org. Chem.* **1998**, *63*, 4181–4192.

125. M. Ashwell, W. Clegg, R. F. W. Jackson, *J. Chem. Soc., Perkin Trans. 1,* **1991**, 897–908.

126. A. B. Smith III, M. O. Duffey, *Synlett* **2004**, 1363–1366.

127. G. Majetich, R. Desmond, A. M. Casares, *Tetrahedron Lett.* **1983**, *24*, 1913–1916.

128. B. M. Trost, P. J. Bonk, *J. Am. Chem. Soc.* **1985**, *107*, 1778–1781.

129. S. D. Rychnovsky, D. Fryszman, U. R. Khire, *Tetrahedron Lett.* **1999**, *40*, 41–44.

130. J.-F. Margathe, M. Shipman, S. C. Smith, *Org. Lett.* **2005**, *7*, 4987–4990.

131. K. Mori, *J. Heterocycl. Chem.* **1996**, *33*, 1497–1517.

132. R. W. Hoffmann, B. Kemper, *Tetrahedron Lett.* **1982**, *23*, 845–848.

133. Y. Yamamoto, Y. Saito, K. Maruyama, *Tetrahedron Lett.* **1982**, *23*, 4959–4962.

134. P. G. M. Wuts, S. S. Bigelow, *Synth. Commun.* **1982**, *12*, 779–785.

135. H. H. Wasserman, E. H. Barber, *J. Am. Chem. Soc.* **1969**, *91*, 3674–3675.

136. I. R. Trehan, J. Singh, K. Ajay, J. Kaur, G. L. Kad, *Indian J. Chem.* **1995**, *34B*, 396–398.

Chapter 3
Skeleton Oriented Bond-Sets

Abstract Branches in the target structure mark points at which bonds should be made during synthesis. If no functional group is close to the branching point, an auxiliary functional group has to be introduced temporarily in order to allow construction of the desired skeletal bond. A substantial reduction in the number of construction steps may be realized, if the target structure or an intermediate has c_2 or σ-symmetry.

Only in rare cases does the target molecule possess an unbranched linear molecular skeleton. More often than not, one is faced with target structures that display branched chains, rings, and substituted rings. In a synthesis, unless branches come with the starting materials, they result from bond formation. This leads to two more approaches worth considering during synthesis planning. In the first, skeleton oriented bond-sets, the bond-set for a molecule with a branched skeleton has to be chosen such that the branches are being formed. Alternatively, the second approach is to move to a building block oriented bond-set, when suitable building blocks containing the required kind of branches are available. Bond-sets following each directive are shown in Scheme 3.1. Both will be discussed below.

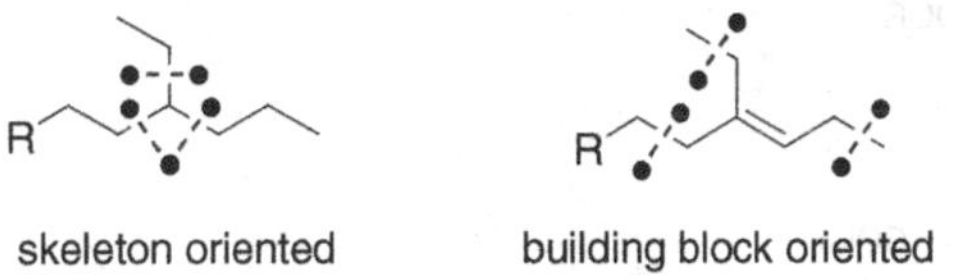

skeleton oriented   building block oriented

Scheme 3.1 Bond-sets with skeleton orientation and with building block orientation

When one deals with molecules having a branched skeleton, one checks the distance between the branching point and any existing functionality. When this distance falls within the normal reach of the functional group, one tends primarily to use natural synthons in order to create a skeletal bond at the branching point (Scheme 3.2).

R.W. Hoffmann, *Elements of Synthesis Planning*,
DOI 10.1007/978-3-540-79220-8_3, © Springer-Verlag Berlin Heidelberg 2009

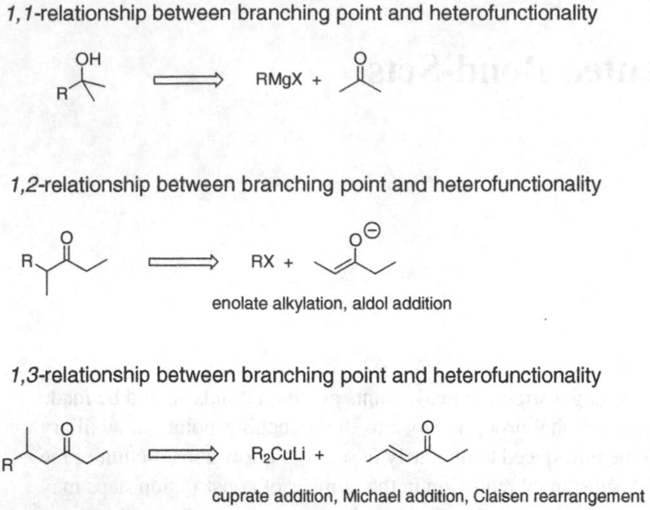

1,1-relationship between branching point and heterofunctionality

1,2-relationship between branching point and heterofunctionality

enolate alkylation, aldol addition

1,3-relationship between branching point and heterofunctionality

cuprate addition, Michael addition, Claisen rearrangement

Scheme 3.2 Bond formation according to the distance betweeen branching point and heteroatom functionality

The available options are exemplified in Scheme 3.3 by a multifunctional-ized but simply branched intermediate **15** taken from the tetracycline synthe-sis of Woodward [1]. Remember to always choose the cut at the branching point.

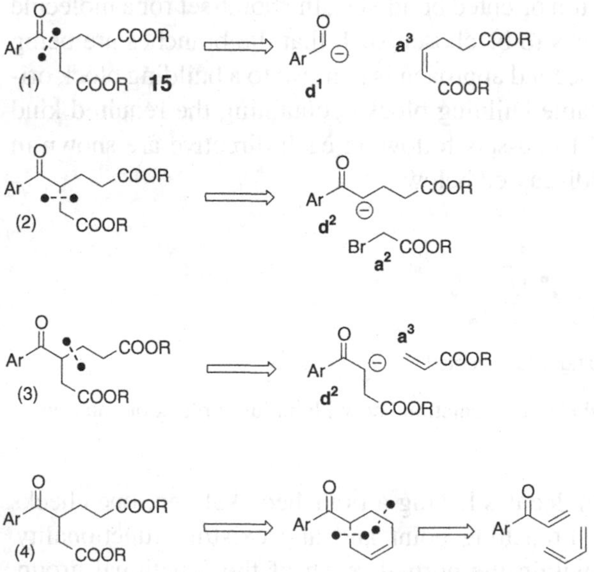

Scheme 3.3 Bond formation at the branching point according to the heterofunctionali-ties present

Of the possibilities shown in Scheme 3.3, (1) is the least attractive since it requires an umpoled d^1-synthon to generate a *1,4*-relationship of functionalities. The approach reflected by (2) is better since it also uses an umpoled, but readily available, a^2-synthon in addition to a natural d^2-synthon. Approach (3) is better still, since it aims at a *1,5*-relationship of functionalities and avoids the use of umpoled synthons. This is the tactic that was employed by Woodward in a modified version [1]. In hindsight, one recognizes an even more attractive route (4) that relies on a *1,6*-relationship of the ester functionalities and a "reconnect" transformation, opening an entry via a Diels-Alder cycloaddition.

When a branching point in the skeleton falls outside the reach of a functional group, one can rely on skeletal bond-forming reactions which do not require the presence of a functional group. A nearly ideal solution to this problem is provided by the transition metal-catalyzed coupling reactions of alkylzinc or alkylmagnesium reagents with alkyl iodides [2, 3, 4, 5] (Scheme 3.4).

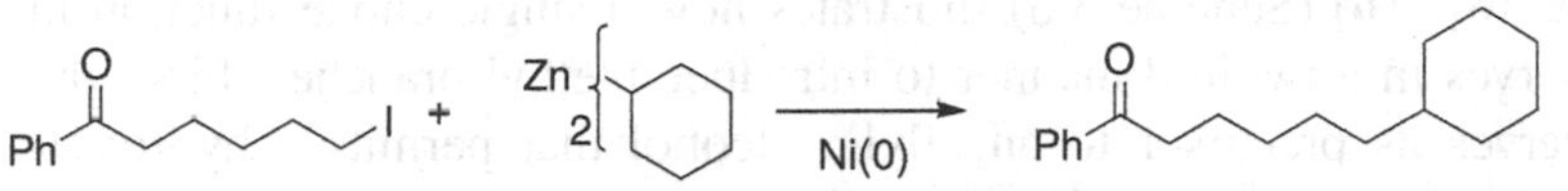

Scheme 3.4 Skeletal bond formation remote from preexisting functionality

3.1 The FGA-Strategy for Preparing Branched Skeletons

On perusal of many natural product syntheses, one notes that detours are frequently taken in order to generate branches in the skeletons. Additional functionality is placed at or close to the point where the bond is to be made. The purpose of this functional group addition (FGA) is to facilitate bond formation at the desired position. This auxiliary functionality has to be removed by an extra step later in the synthesis. In his synthesis of the intermediate **15**, Woodward used a methoxycarbonyl group as an auxiliary functionality [1] (Scheme 3.5).

Of course, introduction and later removal of the methoxycarbonyl group adds two extra steps to the overall synthesis.

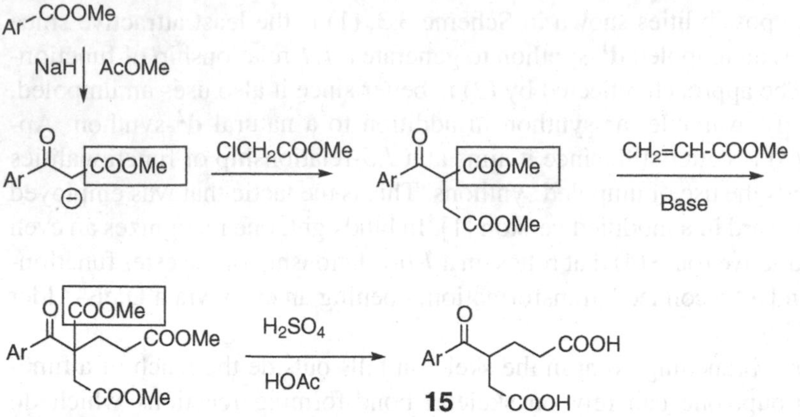

Ar COOMe
NaH AcOMe
COOMe
ClCH2COOMe
COOMe
COOMe
CH2=CH-COOMe
Base
COOMe
COOMe
COOMe
H2SO4
HOAc
COOH
COOH
15

Scheme 3.5 Use of a methoxycarbonyl group as an auxiliary functionality to facilitate bond formation for a branched sekeleton

A standard auxiliary functional group allowing the introduction of branches into a molecular skeleton is the carbonyl group. The synthesis of alnusenone [6] (**16**) (Scheme 3.6) illustrates how a single enone function in ring E serves in a twofold manner to introduce methyl branches. First, the enone serves as precursor to an allylic alcohol that permits a hydroxyl-directed Simmons-Smith cyclopropanation to eventually generate a methyl branch in the ß-position, a tactic which capitalizes on the equivalency of a carbonyl and an alcohol function in retrosynthetic analysis. Second, after reoxidation to a ketone, the carbonyl group allows two consecutive enolate alkylations to introduce two methyl branches directly in the α'-position. Finally, after having orchestrated all these branch-forming steps, the carbonyl group is reductively removed.

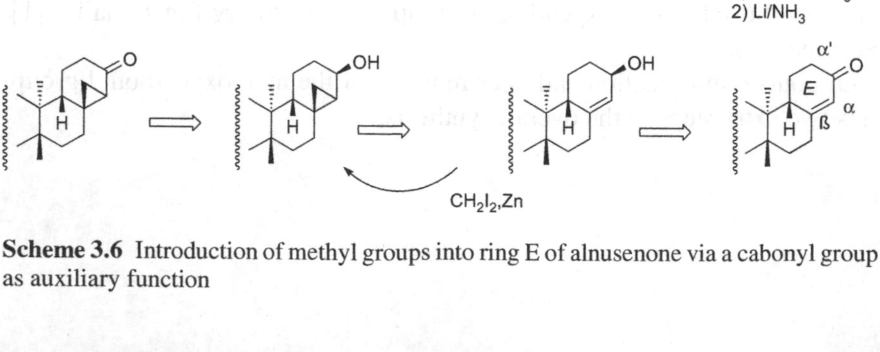

16
FGA
Wolff-Kishner
1) KOtBu, CH3I
2) Li/NH3
OH
OH
CH2I2,Zn
E
α'
α
ß

Scheme 3.6 Introduction of methyl groups into ring E of alnusenone via a cabonyl group as auxiliary function

The utility of a carbonyl group as an auxiliary to introduce branches in a skeleton is underscored by a suggested synthesis of the insect pheromone **17** [7] (Scheme 3.7). The plan of this synthesis is clearly skeleton oriented.

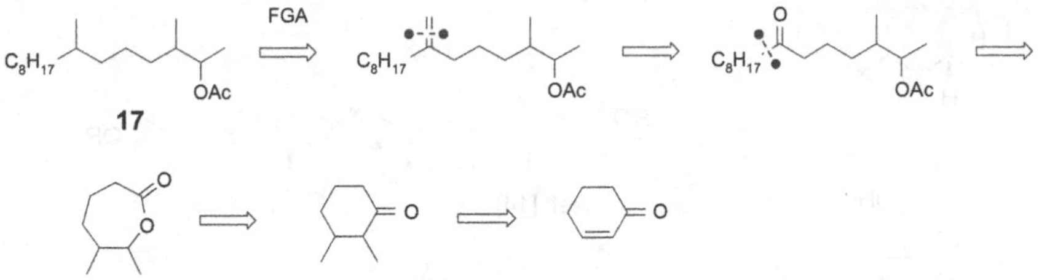

Scheme 3.7 Carbonyl group as auxiliary functionality to generate branches in a molecular skeleton

More recently, arylsulfonyl groups have found use as auxiliary functionalities to allow access to branches in molecular skeletons. Alkylation of an α-sulfonylalkyllithium species such as **18** is quite useful to make skeletal bonds remote from any other controlling functionality. An example is given by the synthesis of diumycinol [8]. In this case, the auxiliary sulfonyl group is disposed of in a skeleton and branch forming Julia–Lythgoe olefination (Scheme 3.8).

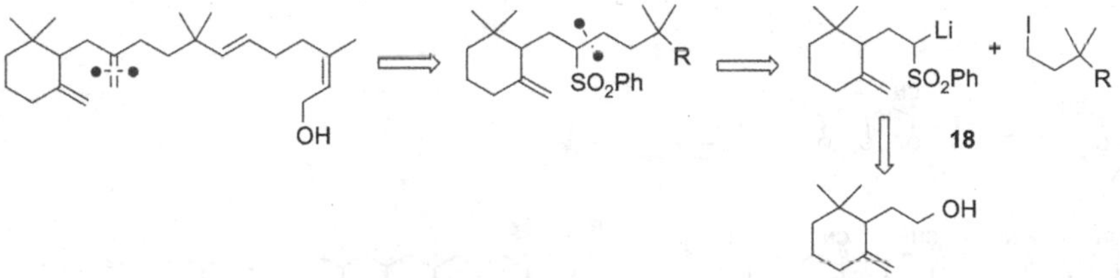

Scheme 3.8 Sulfonyl group-mediated access to a branching point during a synthesis of diumycinol

The attractivness of a sulfonyl group as an auxiliary function in building molecular skeletons is enhanced by its ease of removal. It may serve as a precursor for Julia–Lythgoe olefination [9], or it can be removed reductively under mild conditions [10, 11]. Several cases of bond formation remote from controlling functionality aided by a sulfonyl group are summarized in Scheme 3.9.

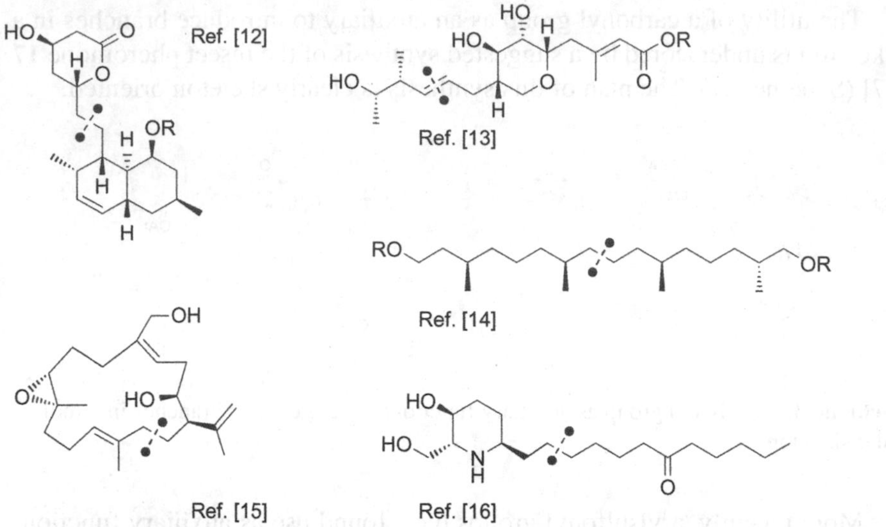

Scheme 3.9 Target structures whose syntheses rely upon sulfonyl group mediated bond formation

The fact that the utilization of a sulfonyl group generally requires two extra steps does not appear to detract from its popularity. Nevertheless, there are other functional groups such as the triphenylphosphonium moiety or nitrile groups that may serve in exactly the same manner [17] (Scheme 3.10).

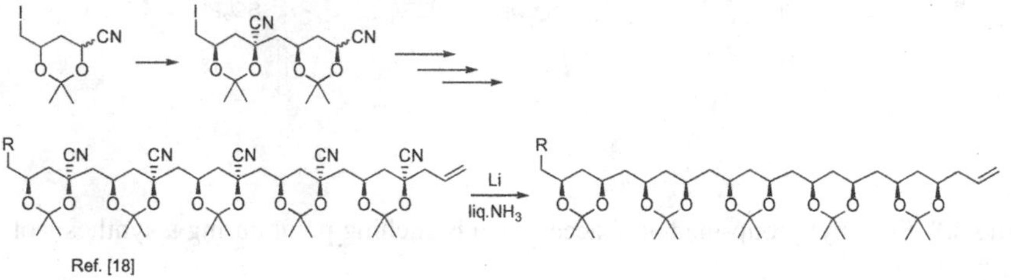

Scheme 3.10 Use of nitrile groups as auxiliary functions for the formation of molecular skeletons

Nitrile groups may be be removed reductively either with LiDBB [19] or with Li in liquid ammonia [18, cf. also 20]. Conditions for nitrile removal are not as mild, however, as those required to remove sulfonyl groups, which explains the popularity of the latter.

Another auxiliary functionality which can serve well for making bonds remote from a controlling functionality is a carbon-carbon double bond. A double bond facilitates bond formation in its vicinity and may in the end be removed by catalytic hydrogenation. Considering target molecule **19,** the following retrosynthetic analysis is suggested (Scheme 3.11).

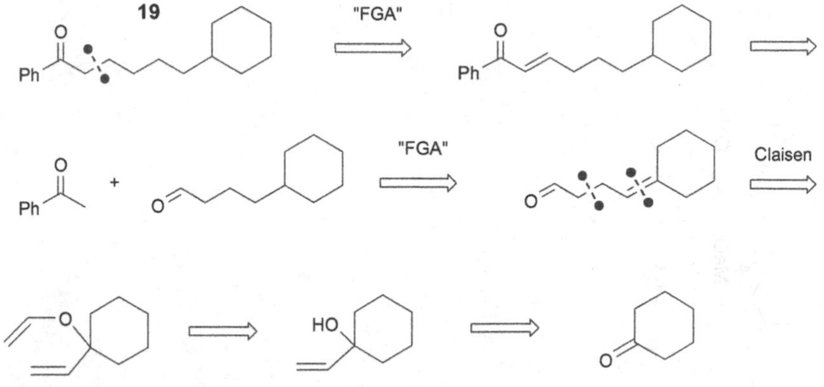

Scheme 3.11 Use of a carbon-carbon double bond as an auxiliary function to allow the introduction of a branch far from a preexisting functionality

Bond formation remote from functional groups is frequently required when following a building block oriented approach to a molecular skeleton. On considering a synthesis of cylindrocyclophane [21] **(20)** (cf. Scheme 3.12), the symmetry of the target suggests a dimerization of identical building blocks. This should give rise to a macrocycle indicating olefin metathesis as the key reaction. Therefore an olefinic double bond becomes the auxiliary structural element to enable linkage of the two units.

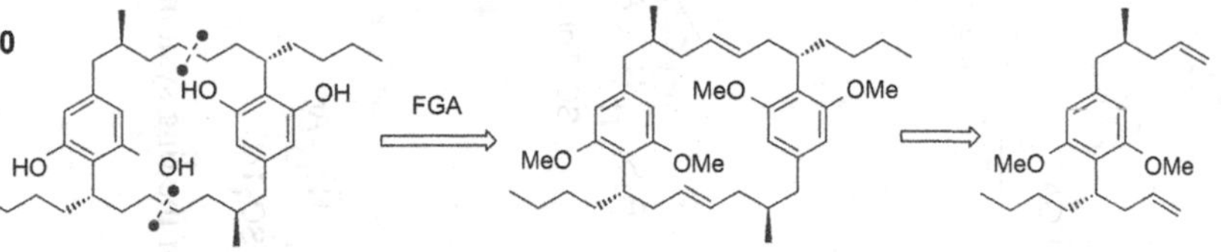

Scheme 3.12 Concept of the synthesis of cylindrocyclophane using carbon-carbon double bonds as auxiliary groups to effect macrocyclzation

The intended macrocyclization is threatened, in theory, by a regioselectivity problem. In practice, the ring-closing metathesis proceeded with a high regioselectivity in favor of the desired head-to-tail dimerization [21]. Since this was not clear at the outset, the exploration of the synthesis route was initiated by a stepwise linkage of the building blocks (Scheme 3.13) to ascertain the correct regioselectivity in the overall process [22].

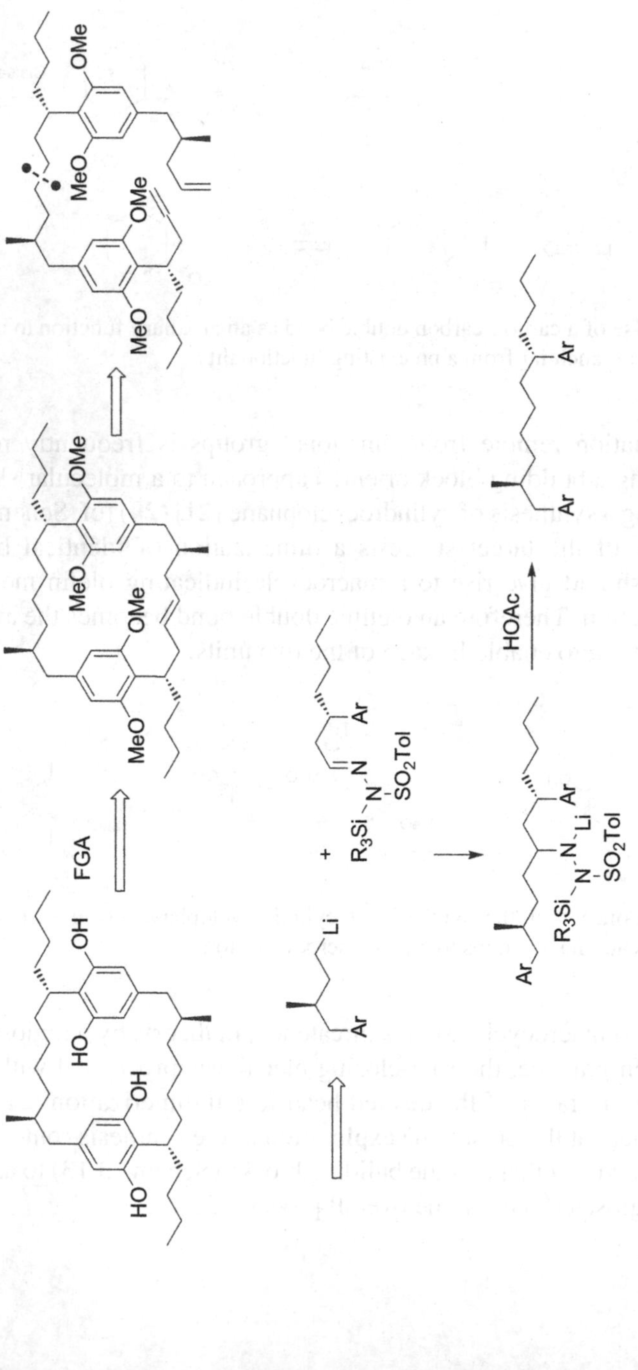

Scheme 3.13 Stepwise formation of the ring system of cyclindrocyclophane

For these stepwise carbon-carbon bond-forming reactions a tosylhydrazone was employed as an auxiliary function [23]. The tosylhydrazine is directly removed from the product by acidic hydrolysis, after it has done its job in enabling skeletal bond formation.

The temporary presence of a carbon-carbon double bond can be advantageous for creating branches in a molecular skeleton by bond formation. This functionality can be obtained not only by olefin metathesis, but also by any carbonyl olefination reaction (Wittig [24], Horner-Wadsworth-Emmons [25], Peterson, or Julia-Lythgoe [9]) (Scheme 3.14).

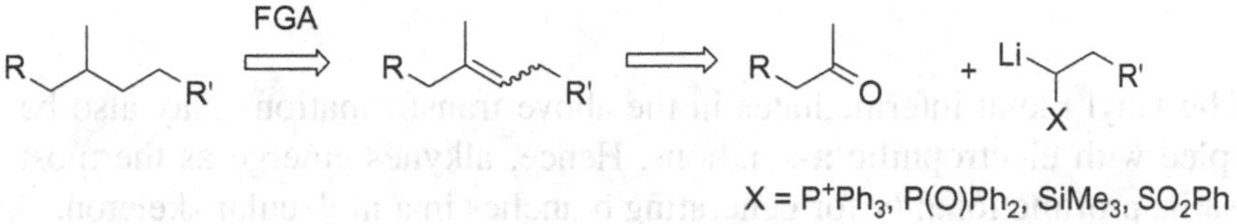

$X = P^+Ph_3, P(O)Ph_2, SiMe_3, SO_2Ph$

Scheme 3.14 Approach to branches in a skeleton based on carbonyl olefination reactions

When forging skeletal bonds near an olefin, the formation of vinylic bonds has assumed a prominent role due to the development of the transition metal-catalyzed coupling reactions (Scheme 3.15).

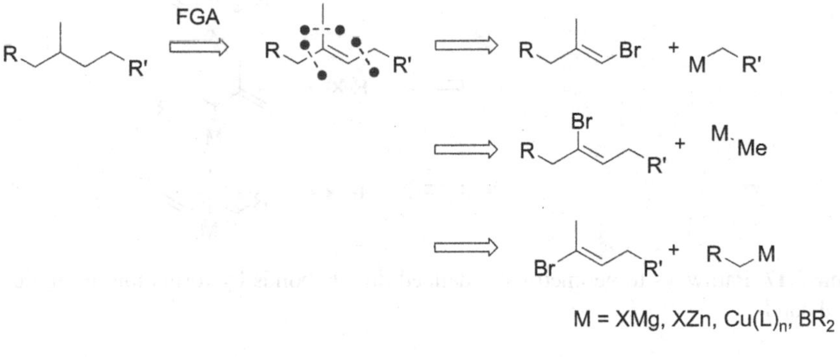

$M = XMg, XZn, Cu(L)_n, BR_2$

Scheme 3.15 Introduction of branches into a skeleton by formation of vinylic bonds based on a carbon-carbon double bond as an auxiliary function

For successful application of this strategy, ready access to vinyl halides with defined configuration of the carbon-carbon double bond is required. Such building blocks can be obtained via carbometallation of terminal alkynes [26] or by hydrometallation of internal alkynes, often with appropriate regioselectivity [27] (Scheme 3.16).

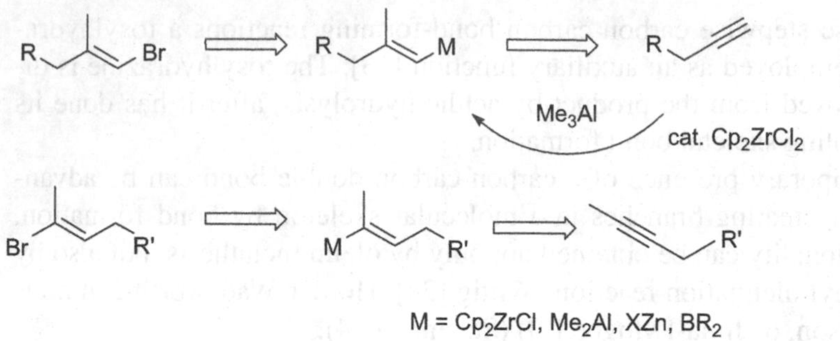

$$M = Cp_2ZrCl, Me_2Al, XZn, BR_2$$

Scheme 3.16 Pathways to vinyl halides with defined configuration of the double bond

The vinyl metal intermediates in the above transformations may also be coupled with electrophilic **a**-synthons. Hence, alkynes emerge as the most versatile profunctionality for generating branches in a molecular skeleton.

The formation of bonds allylic to a carbon-carbon double bond is by no means less developed (Scheme 3.17).

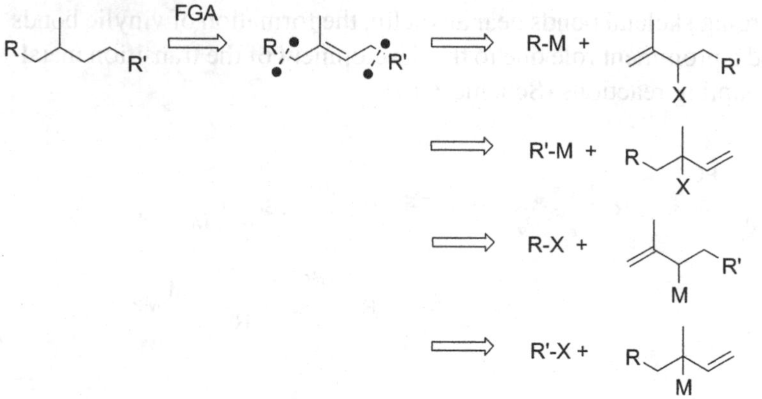

Scheme 3.17 Pathways to geometrically defined double bonds by formation of allylic skeletal bonds

When skeletal bonds to groups R and R' are to be made in the vicinity of a carbon-carbon double bond, substitution of allylic acetates by organocuprates or by malonates (Pd(0)-catalyzed) comes to mind. Alternatively, when R and R' are to be introduced as electrophiles (**a**-synthons), one considers the Lewis acid catalyzed substitution of allylsilanes or allylstannanes. The incorporation of a branched methallyl or isoprene moiety in this manner represents a standard example of a building block oriented strategy in synthesis [28] (Scheme 3.18).

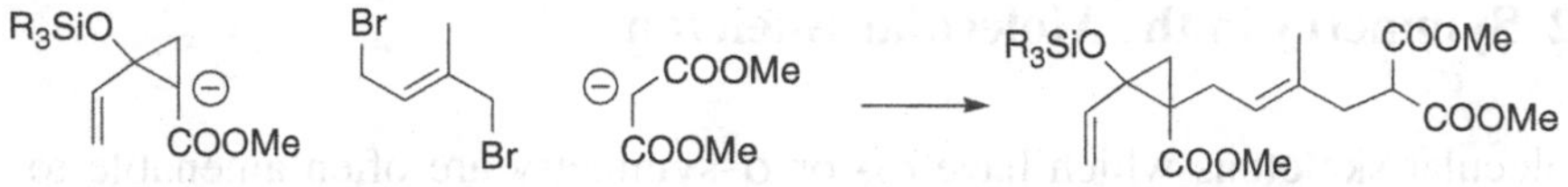

Scheme 3.18 Incorporation of a branched building block into a skeleton

The generation of allylic bonds with (or without) concomitant formation of skeletal branching may be readily achieved via sigmatropic rearrangements such as Cope or Claisen rearrangements (Scheme 3.19). These occupy a favored position among the carbon-carbon bond forming reactions, because they allow control of stereochemistry as well.

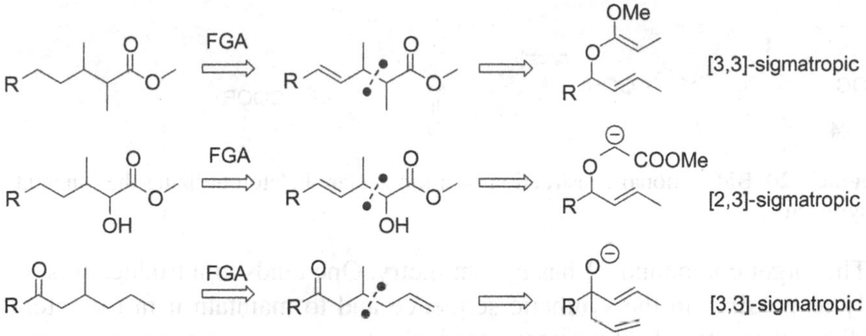

Scheme 3.19 Formation of allylic bonds and branches in one step by sigmatropic rearrangments

In summary: In a target structure with branches in the skeleton the retrosynthetic cuts should be placed at the branching point, in order to generate the branches in the forward synthesis. As there are usually three options to make such a cut, it is done with regard for the existing functionality. If, however, the branching point is outside the reach of the existing functionality, one tends to introduce auxiliary functional groups (i.e., FGA). The most versatile of such auxiliary functions is the carbon-carbon double bond, provided it can later be removed by hydrogenation without impacting other reducible functional groups. An alternative auxiliary functional group is the (readily removable) arylsulfonyl group. Finally, carbonyl groups meet many of the criteria for assisting in bond formation, and hence branch formation. However, their ultimate removal frequently requires a series of reaction steps that may involve quite harsh conditions or may not be feasible at all [29].

3.2 Symmetry in the Molecular Skeleton

Molecular skeletons which have c_2- or σ-symmetry are often amenable to efficient syntheses. The number of necessary synthetic steps can be reduced when one succeeds in making two "symmetrical" skeletal bonds simultaneously. This is illustrated with respect to the synthesis of compound **21** [30] in Scheme 3.20.

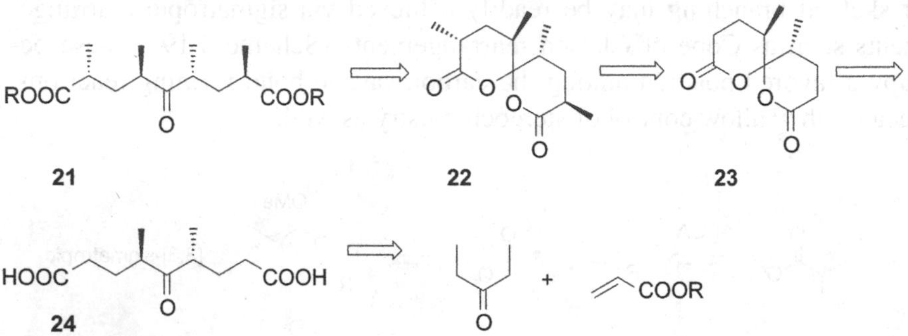

Scheme 3.20 Bidirectional construction of a molecular skeleton utilizing the inherent c_2-symmetry

The target compound **21** has c_2-symmetry. One tends to introduce this as early as possible in the synthetic sequence and to maintain it in the intermediates thereafter. In the present example, the c_2-symmetry prevails from the intermediate **24** through compounds **23** and **22** to the target **21**. Intermediate **24** is generated in a symmetrical fashion by double Michael addition of diethyl ketone to methyl acrylate. Subsequent double methylation of **23** provides **22**, retaining c_2-symmetry. This illustrates how the utilization of symmetry allows for a bidirectional elaboration of the molecular skeleton [31, 32], by which two skeletal bonds are formed at a time. The advantage gained in this fashion is cause for attempting to reduce a nonsymmetrical target molecule to a symmetrical precursor molecule, if possible. This is nicely demonstrated by the synthesis presented [33] in Scheme 3.21.

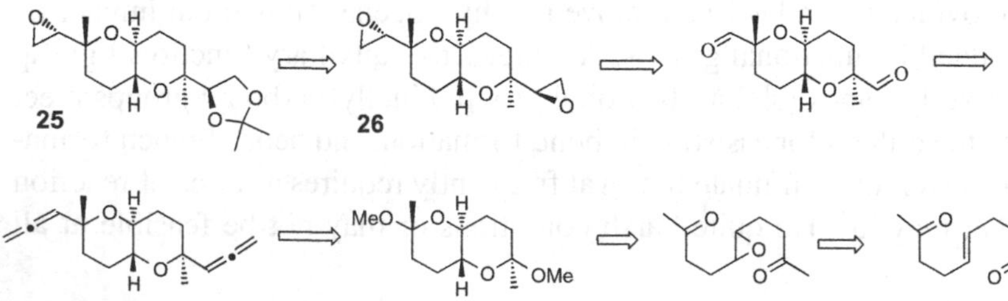

Scheme 3.21 Reduction of a nonsymmetrical target structure to an intermediate with *i*-symmetry to allow a bidirectional synthetic strategy

In this example, compound **26,** having a center of inversion, was identified as a potential precursor to the nonsymmetrical target structure **25.** Compound **26** was then prepared using all the advantages of bidirectional synthesis. Symmetry was broken only in the final stage of the synthesis. Intermediate **26** is a meso compound; and so the two epoxide groups in **26** are enantiotopic to one another. Hence, the reagent to effect desymmetrization must be enantiomerically pure, in order to differentiate the enantiotopic ends in a sort of kinetic resolution-like sense. This was achieved by enantioselective epoxide hydrolysis (Jacobsen reaction) [34]. Importantly, this symmetry-driven synthesis of **25** was much more efficient than an earlier synthesis [35], which did not capitalize on the latent symmetry of the target structure.

As previously indicated for the example in Scheme 3.20, molecule **21** has c_2-symmetry. The ends of this molecule are homotopic to one another. Desymmetrization of such a molecule, if desired, can be achieved by changing one end of the molecule in any manner. Because the two ends of the molecule are identical, it does not matter which end is changed, as long as only one end is changed. In other words, desymmetrization of a c_2-symmetrical intermediate is easier than desymmetrization of a meso compound. Either of these symmetry elements, if present in a molecule, opens up the possibility for one to employ a bidirectional construction strategy. Yet the bidirectional elaboration of a meso compound is more difficult than that of c_2-symmetrical compounds [36]. However, it may be difficult to recognize latent symmetry in complex target structures [37]. Consider the example of neohalicholactone (**27**) [38] given in Scheme 3.22.

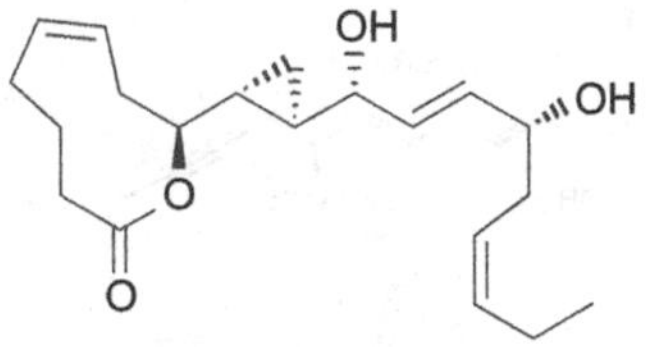

27

Scheme 3.22 Neohalicholactone, a molecule with latent symmetry?

By unravelling the lactone ring, one is better able to appreciate the symmetry embedded in neohalicholactone. Thus, an acyclic meso compound emerges as a suitable precursor to the target structure. After further inspection, a c_2-symmetrical precursor is conceived simply by inverting the configuration at one of the hydroxyl-bearing carbon atoms and eliminating the stereogenic center in the middle of the molecule (Scheme 3.23).

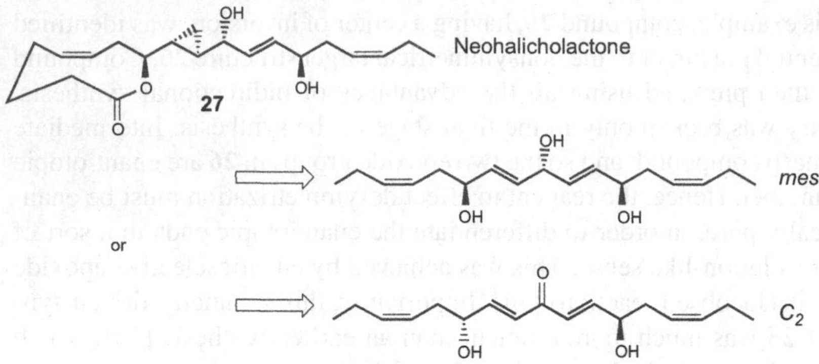

Scheme 3.23 Potential symmetrical precursor structures for neohalicholactone

Consider the c_2-symmetrical precursor **28**, in which both acetoxy groups are homotopic (identical). Desymmetrization by hydrolysis of just one acetyl residue could be readily achieved late in a projected synthesis. This would enable a hydroxyl-directed Simmons-Smith cyclopropanation. Mitsunobu esterification (with inversion of configuration) would then set the stage for a concluding alkyne metathesis [39] (Scheme 3.24). This approach does not yet address the formation of the stereogenic center in the middle of the target structure, be it by oxidation to a ketone and stereoselective reduction.

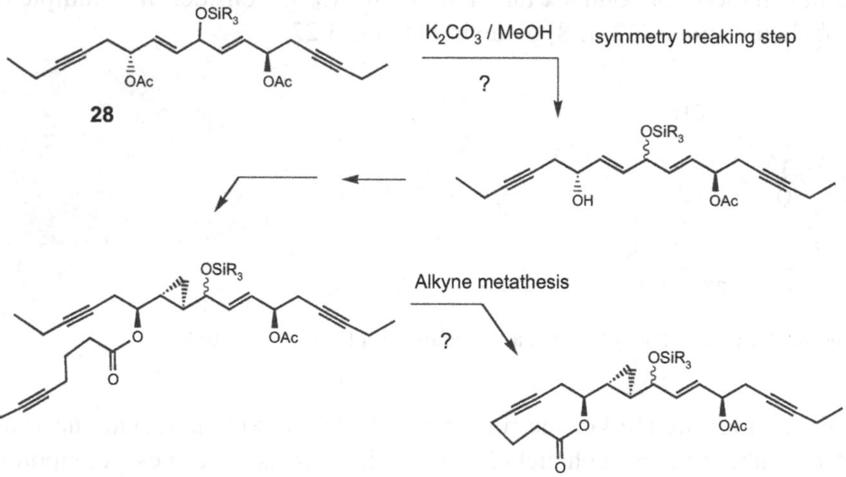

Scheme 3.24 Proposed synthesis for neohalicholactone utilizing a c_2-symmetrical intermediate

It is equally possible to derive a synthesis plan for neohalicholactone (**27**) via the *meso* precursor shown in Scheme 3.23. But when both options are available, one should first consider an approach involving a c_2-symmetrical precursor [36].

Problems

3.1 Integerrinecic acid (Scheme 3.25), despite its small size, has branches and functional groups enough to practise meaningful retrosynthesis. The construction plan of three very similar syntheses [40, 41, 42] (sequence of bond formation (1), (2), (3)) shows that all cuts are made to create branches with the aid of the existent functionality. Work backwards (3) → (2) → (1) to recognize by which reactions a synthesis can be realized. For a completely different approach, see reference [43].

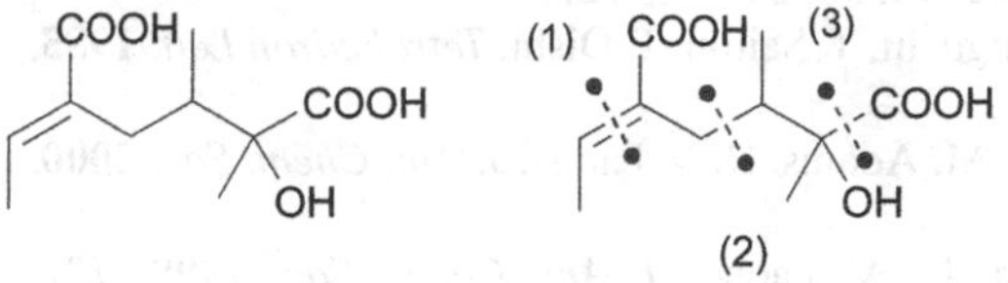

Scheme 3.25 Integerrinecic acid and bond-set for synthesis

3.2 Look for symmetrical building blocks as potential precursors to the following compounds (Scheme 3.26).

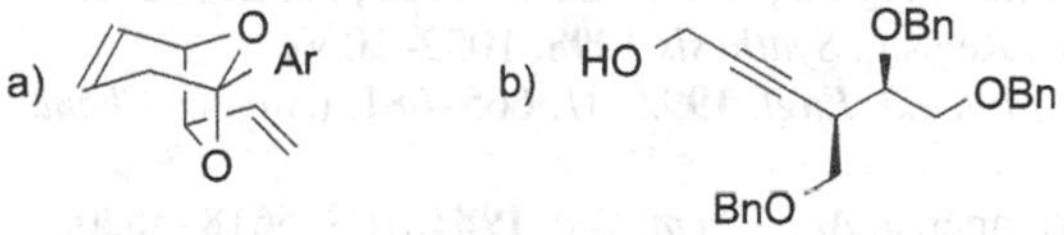

Scheme 3.26 Target molecules incorporating hidden symmetry

References

1. J. J. Korst, J. D. Johnston, K. Butler, E. J. Bianco, L. H. Conover, R. B. Woodward, *J. Am. Chem. Soc.* **1968**, *90*, 439–457.
2. A. E. Jensen, P. Knochel, *J. Org. Chem.* **2002**, *67*, 79–85.
3. J. Zhou, G. C. Fu, *J. Am. Chem. Soc.* **2003**, *125*, 14726–14727.
4. N. Hadei, E. A. B. Kantchev, C. J. O'Brien, M. G. Organ, *Org. Lett.* **2005**, *7*, 3805–3807.

5. C. Herber, B. Breit, *Angew. Chem., Int. Ed.* **2005**, *44*, 5267–5269. (*Angew. Chem.* **2005**, *117*, 5401–5403).
6. R. E. Ireland, M. I. Dawson, S. C. Welch, A. Hagenbach, J. Bordner, B. Trus, *J. Am. Chem. Soc.* **1973**, *95*, 7829–7841.
7. G. Magnusson, *Tetrahedron* **1978**, *34*, 1385–1388.
8. P. Kocienski, M. Todd, *J. Chem. Soc., Perkin Trans. 1,* **1983**, 1783–1789.
9. P. J. Kocienski, *Chem. & Ind.* **1981**, 548–551.
10. C. Nájera, M. Yus, *Tetrahedron* **1999**, *55*, 10547–10658.
11. I. Das, T. Pathak, *Org. Lett.* **2006**, *8*, 1303–1306.
12. J. R. Falck, Y.-L. Yang, *Tetrahedron Lett.* **1984**, *25*, 3563–3566.
13. G. E. Keck, D. F. Kachensky, E. J. Enholm, *J. Org. Chem.* **1984**, *49*, 1462–1464.
14. C. H. Heathcock, P. A. Radel, *J. Org. Chem.* **1986**, *51*, 4322–4323.
15. M. A. Tius, A. Fauq, *J. Am. Chem. Soc.* **1986**, *108*, 6389–6391.
16. T. N. Birkinshaw, A. B. Holmes, *Tetrahedron Lett.* **1987**, *28*, 813–816.
17. Y. Fall, M. Torneiro, L. Castedo, A. Mourino, *Tetrahedron Lett.* **1992**, *33*, 6683–6686.
18. S. D. Rychnovsky, G. Griesgraber, *J. Org. Chem.* **1992**, *57*, 1559–1563.
19. D. Guijarro, M. Yus, *Tetrahedron* **1994**, *50*, 3447–3452.
20. T. Ohsawa, T. Kobayashi, Y. Mizuguchi, T. Saitoh, T. Oishi, *Tetrahedron Lett.* **1985**, *26*, 6103–6106.
21. A. B. Smith III, S. A. Kozmin, C. M. Adams, D. V. Paone, *J. Am. Chem. Soc.* **2000**, *122*, 4984–4985.
22. A. B. Smith III, S. A. Kozmin, D. V. Paone, *J. Am. Chem. Soc.* **1999**, *121*, 7423–7424.
23. A. G. Myers, M. Movassaghi, *J. Am. Chem. Soc.* **1998**, *120*, 8891–8892.
24. B. E. Maryanoff, A. B. Reitz, *Chem. Rev.* **1989**, *89*, 863–927.
25. J. Clayden, S. Warren, *Angew. Chem., Int. Ed. Engl.* **1996**, *35*, 241–270. (*Angew. Chem.* **1996**, *108*, 261–291).
26. P. Wipf, S. Lim, *Angew. Chem., Int. Ed. Engl.* **1993**, *32*, 1068–1071. (*Angew. Chem.* **1993**, *105*, 1095–1097).
27. Y. Gao, K. Harada, T. Hata, H. Urabe, F. Sato, *J. Org. Chem.* **1995**, *60*, 290–291.
28. A. Ullmann, J. Schnaubelt, H.-U. Reissig, *Synthesis* **1998**, 1052–1066.
29. C. H. Heathcock, *Angew. Chem., Int. Ed. Engl.* **1992**, *31*, 665–681. (*Angew. Chem.* **1992**, *104*, 675–691).
30. T. R. Hoye, D. R. Peck, P. K. Trumper, *J. Am. Chem. Soc.* **1981**, *103*, 5618–5620.
31. C. S. Poss, S. L. Schreiber, *Acc. Chem. Res.* **1994**, *27*, 9–17.
32. S. R. Magnuson, *Tetrahedron* **1995**, *51*, 2167–2213.
33. J. M. Holland, M. Lewis, A. Nelson, *Angew. Chem., Int. Ed.* **2001**, *40*, 4082–4084. (*Angew. Chem.* **2001**, *113*, 4206–4208).
34. M. Tokunaga, J. F. Larrow, F. Kakiuchi, E. N. Jacobsen, *Science* **1997**, *277*, 936–938.
35. T. Nakata, *J. Synth. Org. Chem., Jpn.* **1998**, *56*, 940–951.
36. R. W. Hoffmann, *Angew. Chem., Int. Ed.* **2003**, *42*, 1096–1109. (*Angew. Chem.* **2003**, *115*, 1128–1142).
37. M. Ball, M. J. Gaunt, D. F. Hook, A. S. Jessiman, S. Kawahara, P. Orsini, A. Scolaro, A. C. Talbot, H. R. Tanner, S. Yamanoi, S. V. Ley, *Angew. Chem., Int. Ed.* **2005**, *44*, 5433–5438. (*Angew. Chem.* **2005**, *117*, 5569–5574).
38. D. J. Critcher, S. Connolly, M. Wills, *J. Org. Chem.* **1997**, *62*, 6638–6657.
39. A. Fürstner, P. W. Davies, *Chem. Commun.* **2005**, 2307–2320.

40. S. E. Drewes, N. D. Emslie, *J. Chem. Soc., Perkin Trans. 1,* **1982**, 2079–2083.
41. U. Pastewka, H. Wiedenfeld, E. Röder, *Arch.Pharm.* **1980**, *313*, 846–850.
42. C.C. J. Culvenor, T. A. Geissman, *J. Am. Chem. Soc.* **1961**, *83*, 1647–1652.
43. K. Narasaka, T. Uchimaru, *Chem. Lett.* **1982**, 57–58.

Chapter 4
Building Block Oriented Synthesis

Abstract If substructures with special features (branches, stereogenic centers) of the target correspond to readily available starting materials, it is advisable to incorporate those as building blocks in the synthesis. Guidelines are given as to how to identify suitable building blocks.

One tends to pursue a building block oriented synthesis when building blocks are available that contain characteristic structural elements present in the target structure. Frequently, such structural elements are stereochemistry related, e.g., the defined configuration of a multiply- substituted double bond or a certain sequence of contiguous stereogenic centers. When the synthesis of compound **29** (the cecropia juvenile hormone) was considered, the thiapyrane **30** was identified as a suitable precursor, since this subunit contains the appropriate number of carbon atoms along with the correct double bond configuration [1, 2] (Scheme 4.1).

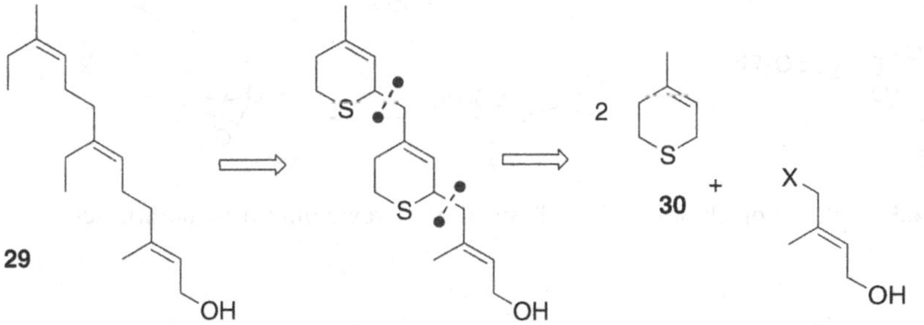

Scheme 4.1 Identification of a building block containing the correct double bond configuration

R.W. Hoffmann, *Elements of Synthesis Planning*,
DOI 10.1007/978-3-540-79220-8_4, © Springer-Verlag Berlin Heidelberg 2009

When the methodology of stereoselective synthesis was still in its infancy, it was considered advantageous to utilize sequences of stereogenic centers available from enantiomerically pure natural products as building blocks [3, 4]; this so-called chiral pool synthesis strategy is exemplified in Scheme 4.2. The bicyclic acetal structure of exo-brevicomin (**31**) can be retrosynthetically linked to the chiral ketodiol **32**, which can be derived from (S,S)-(−)-tartaric acid, a readily available chiral starting material. This leads to the building block oriented bond-set depicted in intermediate **32**.

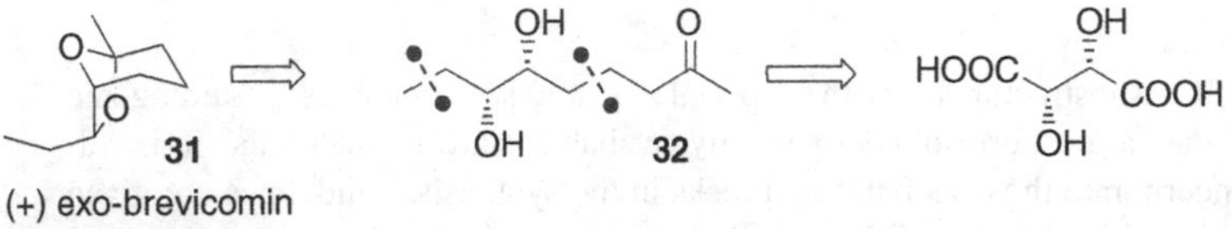

Scheme 4.2 Building block oriented (ex chiral pool) retrosynthesis of exo-brevicomin

Several syntheses of exo-brevicomin have been executed according to this bond-set [5, 6, 7, 8, 9]. Their step count varies between 7 and 12, illustrating that, for a given bond-set, there is still ample room for intelligent planning of a synthesis in the forward direction. One [9] of these syntheses is illustrated in Scheme 4.3.

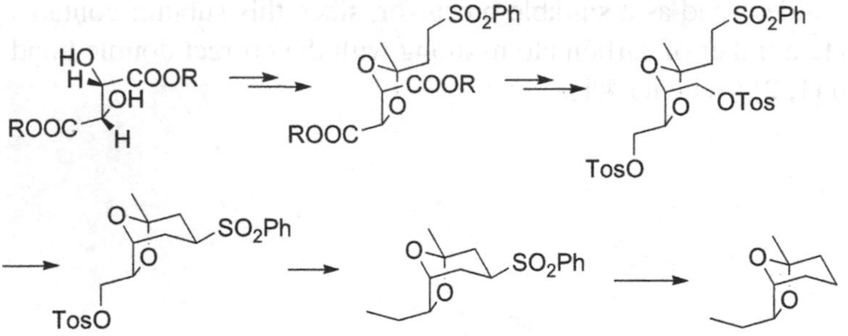

Scheme 4.3 Building block oriented synthesis of exo-brevicomin from tartaric acid

This synthesis uses an auxiliary sulfonyl group (FGA, see Sect. 3.1) to enable the formation of one of the skeletal bonds.

The choice of a suitable chiral precursor is often obvious for a given target structure. However, the obvious choice is not necessarily the only meaningful or possible solution. In the case of eleutherobin **33,** one tends to immediately envision (+)-carvone as a suitable chiral precursor [10]. However, a different adaptation reveals that (−)-carvone could also be an attractive precursor [11]. Even α-phellandrene has been chosen as the starting point for an efficient synthesis of eleutherobin [12] (Scheme 4.4).

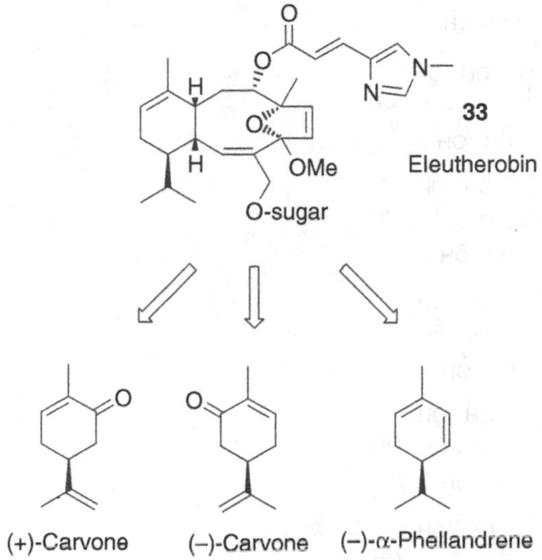

Scheme 4.4 Suitable chiral building blocks for the synthesis of eleutherobin

In order to make the optimal choice from among suitable chiral precursors, one needs a compilation of all available chiral natural products. A selection of these is published in a review by Scott [13]. However, because one tends to write a target structure in a distinct arrangement, and the potential chiral precursors are often depicted quite differently, it can be difficult to recognize similarities or differences in constitution and configuration between target and precursor structures. Such comparisons can be effected reliably by computer programs [14]. Yet when one writes both target structure and precursor structures in the same spatial arrangement, even pedestrian solutions become readily apparent. This is illustrated by a list of common sugar building blocks, written in a zig-zag arrangement of the backbone, from C-6 to C-1 and also in the opposite sense (Schemes 4.5 and 4.6).

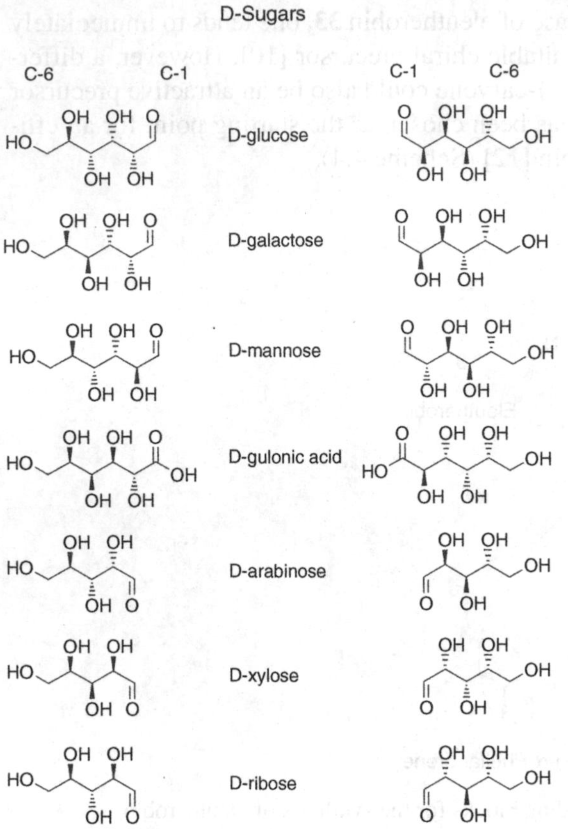

D-Sugars

C-6   C-1   C-1   C-6

OH OH O   O OH OH
HO   D-glucose   OH
   OH OH   OH OH

OH OH O   O OH OH
HO   D-galactose   OH
   OH OH   OH OH

OH OH O   O OH OH
HO   D-mannose   OH
   OH OH   OH OH

OH OH O   O OH OH
HO   OH   D-gulonic acid   HO   OH
   OH OH   OH OH

OH OH   OH OH
HO   D-arabinose   OH
   OH O   O   OH

OH OH   OH OH
HO   D-xylose   OH
   OH O   O   OH

OH OH   OH OH
HO   D-ribose   OH
   OH O   O   OH

Scheme 4.5 Readily available D-sugars in zig-zag arrangement of the main skeleton

L-sugars

C-6   C-1   C-1   C-6

OH O O   O O OH
HO   OH   L-ascorbic acid   HO   OH
   OH O   O   OH

OH OH   OH OH
HO   L-arabinose   OH
   OH O   O   OH

OH OH OH   HO OH OH
HO   L-sorbose   OH
   OH O   O   OH

OH OH O   O OH OH
                     L-rhamnose
   OH OH   HO OH

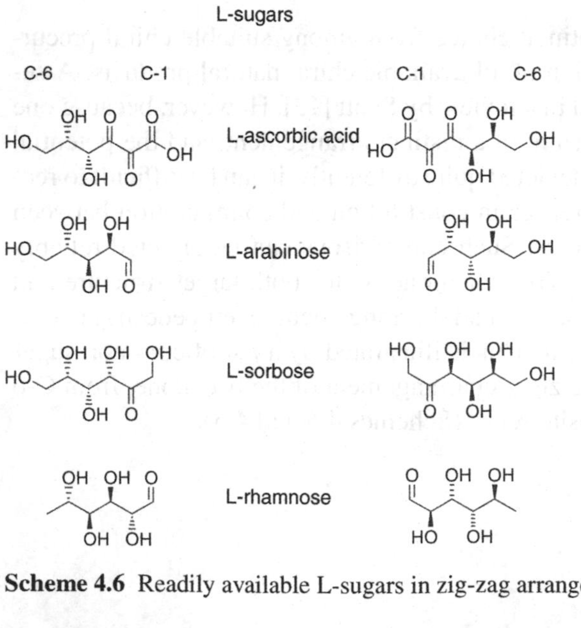

Scheme 4.6 Readily available L-sugars in zig-zag arrangement of the main skeleton

It is advisable to copy these schemes as a transparency. When a target
structure has several oxygenated stereogenic centers along its main chain,
one should write the target structure in a zig-zag arrangement of the main
chain. Then it will be possible by an overlay of the transparency to check
which readily available sugar molecules possess a complete or partial congruence regarding the stereogenic centers. For example, consider the arachi-
donic acid derivative **34**. The comparison shown in Scheme 4.7 indicates that
D-glucose could be a useful precursor. A synthesis along these lines would
require deoxygenation at C-3 of glucose, as well as chain extensions at C-1
and C-6. In fact, an efficient synthesis of compound **34** was accomplished
via this strategy [15].

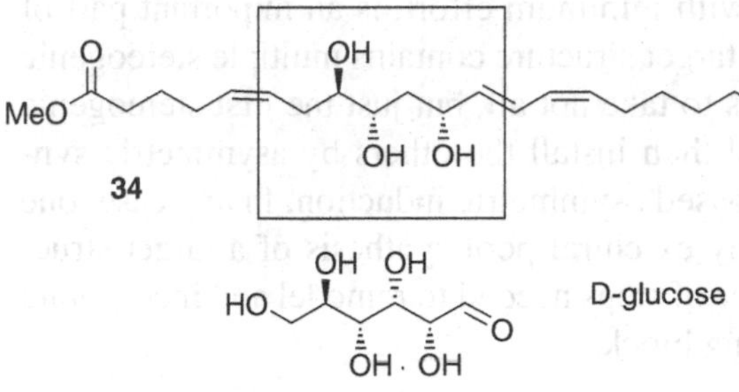

Scheme 4.7 Identification of D-glucose as a suitable precursor for synthesis of **34**

During a synthesis of erythronolide A, carried out by our group at
Marburg, we needed the chiral aldehyde **35** as starting material. Perusal of
the list of commercially available chiral starting materials [13] suggested a
synthesis of lactone **36** from D-fructose (Scheme 4.8). With this in mind,
aldehyde **35** was prepared from fructose in eight steps [16].

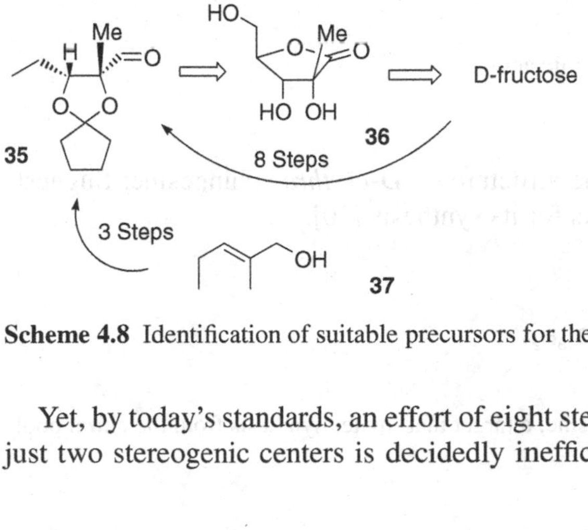

Scheme 4.8 Identification of suitable precursors for the synthesis of **35**

Yet, by today's standards, an effort of eight steps to create a molecule with
just two stereogenic centers is decidedly inefficient! Due to the significant

enhancements in stereoselective synthesis methodology, it is now possible
to access the aldehyde **35** in three steps via Sharpless asymmetric epoxidation beginning with the allylic alcohol **37** [17]. Thus, a principle drawback
of ex chiral pool synthesis is illustrated: an excessive number of steps is required in order to trim down an overfunctionalized natural product during a
synthesis in which it is employed. Ex chiral pool synthesis is only justified
when the chiral building block contains a considerable measure of complexity (e.g., three or more stereogenic centers) that can be incorporated into the
target structure. Long reaction sequences, after which only one stereogenic
remains intact from a complex sugar [18, 19], are justified only if the aim is
to establish absolute configuration by chemical correlation.

The search for suitable chiral precursor molecules, which can be incorporated into a target structure with minimum effort, is an important part of
planning a synthesis. When the target structure contains multiple stereogenic
centers, it may be advantageous to take not all, but just the first stereogenic
center from the chiral pool and then install the others by asymmetric synthesis, preferably by substrate-based asymmetric induction. In any case, one
should think critically about any ex chiral pool synthesis of a target structure, bearing in mind the number of steps needed to remodel and incorporate
a readily available chiral building block.

Problems

4.1 In Scheme 4.9 the core structure of polyoxamic acid is shown. Suggest
suitable chiral building blocks for its synthesis.

$$RO\overset{\underset{\displaystyle HO}{|}}{\diagdown}\overset{\underset{\displaystyle OH}{|}}{\diagup}\overset{\underset{\displaystyle }{NH_2}}{\diagup}COOH \qquad R = H_2N\text{-}CO\text{-}$$

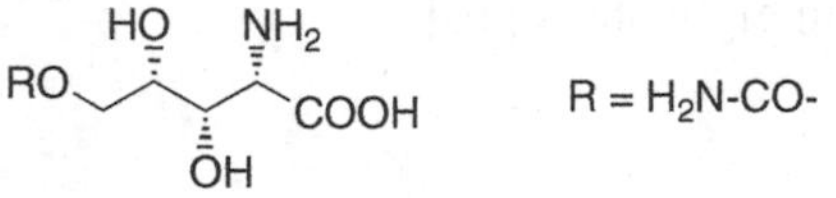

Scheme 4.9 Structure of polyoxamic acid

4.2 Scheme 4.10 displays the structure of D-*erythro*-sphingosine. Suggest
suitable chiral building blocks for its synthesis [20].

$$HO\overset{\underset{\displaystyle OH}{|}}{\diagdown}\overset{\underset{\displaystyle }{NH_2}}{\diagup}\diagup\diagdown R \qquad R = {}^{n}C_{13}H_{27}$$

Scheme 4.10 D-*erythro*-sphingosine, a target that invites synthesis from the chiral pool

References

1. P. L. Stotter, R. E. Hornish, *J. Am. Chem. Soc.* **1973**, *95*, 4444–4446.
2. K. Kondo, A. Negishi, K. Matsui, D. Tunemoto, S. Masamune, *J. Chem. Soc., Chem. Commun.* **1972**, 1311–1312.
3. D. Seebach, H.-O. Kalinowski, *Nachr. Chem. Tech. Lab.* **1976**, *24*, 415–418.
4. S. Hanessian, *Aldrichimica Acta* **1989**, *22*, 3–14.
5. B. Giese, R. Rupaner, *Synthesis* **1988**, 219–221.
6. H. H. Meyer, *Liebigs Ann. Chem.* **1977**, 732–736.
7. K. Mori, Y.-B. Seu, *Liebigs Ann. Chem.* **1986**, 205–209.
8. H. Kotsuki, I. Kadota, M. Ochi, *J. Org. Chem.* **1990**, *55*, 4417–4422.
9. Y. Masaki, K. Nagata, Y. Serizawa, K. Kaji, *Tetrahedron Lett.* **1982**, *23*, 5553–5554.
10. K. C. Nicolaou, T. Ohshima, S. Hosokawa, F. L. van Delft, D. Vourloumis, J. Y. Xu, J. Pfefferkorn, S. Kim, *J. Am. Chem. Soc.* **1998**, *120*, 8674–8680.
11. S. M. Ceccarelli, U. Piarulli, C. Gennari, *Tetrahedron* **2001**, *57*, 8531–8542.
12. X.-T. Chen, C. E. Gutteridge, S. K. Bhattacharya, B. Zhou, T. R. R. Pettus, T. Hascall, S. J. Danishefsky, *Angew. Chem., Int. Ed. Engl.* **1998**, *37*, 185–186. (*Angew. Chem.* **1998**, *110*, 195–197).
13. J. W. Scott in *Asymmetric Synthesis* (Eds.: J. D. Morrison, J. W. Scott), Academic Press, New York, vol. 4, **1984**, pp. 1–226.
14. S. Hanessian, J. Franco, B. Larouche, *Pure. Appl. Chem.* **1990**, *62*, 1887–1910.
15. G. Just, C. Luthe, *Can. J. Chem.* **1980**, *58*, 1799–1805.
16. R. W. Hoffmann, W. Ladner, *Chem. Ber.* **1983**, *116*, 1631–1642.
17. R. Stürmer, *Liebigs Ann. Chem.* **1991**, 311–313.
18. H. Redlich, W. Francke, *Angew. Chem., Int. Ed. Engl.* **1980**, *19*, 630–631. (*Angew. Chem.* **1980**, *92*, 640–641).
19. H. Redlich, J. Xiang-jun, *Liebigs Ann. Chem.* **1982**, 717–722.
20. P. M. Koskinen, A. M. P. Koskinen, *Synthesis* **1998**, 1075–1091.

Chapter 5
The Basis for Planning

Abstract Systematic retrosynthetic analysis of a target structure will point out building blocks and types of reagents needed to carry out the forward synthesis. At this point actual reagents that correspond to the required type have to be identified. This pertains also to the identification of conjunctive reagents needed to combine building blocks.

It should be well appreciated from the preceding chapters that the planning of syntheses is aimed at identifying the key bonds in a target structure, the formation of which would allow the most direct way of attaining the target. During the initial phase of this mental process, the nature of the bond-forming reaction, as well as the polarity type of the bond formation, were not yet specified. Thinking in such a generalized manner relies upon "half-reactions" [1, 2, 3], reactions in which only one partner, (e.g., the dithiane anion) is defined, whereas the other partner (any electrophilic **a**-component, designated in Scheme 5.1 as E-X) remains undefined. Only the type of bond formation is implied (Scheme 5.1). Thinking in terms of half-reactions is an essential element in planning a synthesis.

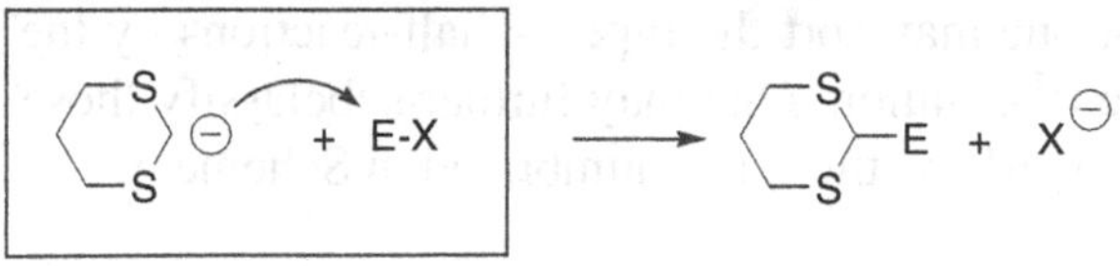

Scheme 5.1 Half-reaction, a reaction between one defined and one undefined partner

The available synthetic reactions can be represented as a combination of two half-reactions each. They can be classified according to certain types of half-reactions [4]. During efforts to arrive at computer-aided synthesis planning, it was recognized that a rather small number of types of half-reactions suffices to encompass the majority of synthetic methodology. Hendrickson

R.W. Hoffmann, *Elements of Synthesis Planning*,
DOI 10.1007/978-3-540-79220-8_5, © Springer-Verlag Berlin Heidelberg 2009

originally defined 29 reaction types, derived from 11 nucleophilic and 4 electrophilic half-reactions [5]. Later this set was expanded to a matrix derived from 16 nucleophilic and 9 electrophilic half-reactions [6]. Some typical half-reactions are illustrated in Scheme 5.2.

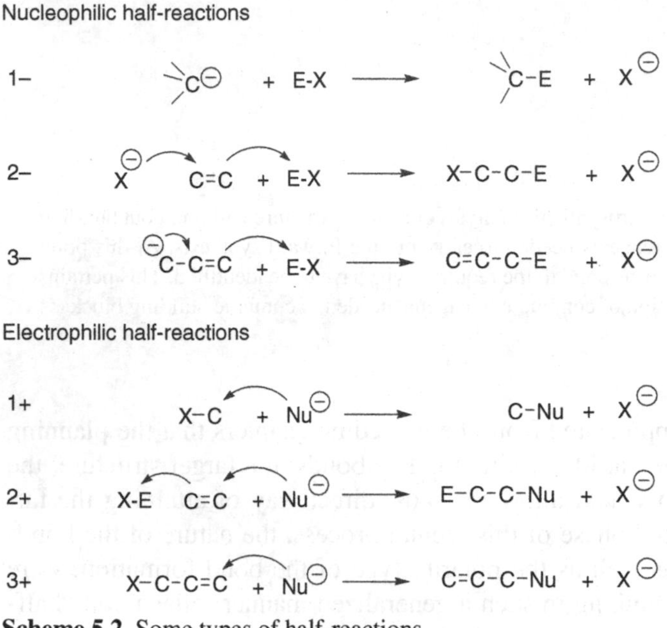

Nucleophilic half-reactions

1− $\quad$ $>\!C^{\ominus}$ + E-X $\longrightarrow$ $>\!C$-E + X$^{\ominus}$

2− $\quad$ X$^{\ominus}$ $\frown$ C=C + E-X $\longrightarrow$ X-C-C-E + X$^{\ominus}$

3− $\quad$ $^{\ominus}$C-C=C + E-X $\longrightarrow$ C=C-C-E + X$^{\ominus}$

Electrophilic half-reactions

1+ $\quad$ X-C + Nu$^{\ominus}$ $\longrightarrow$ C-Nu + X$^{\ominus}$

2+ $\quad$ X-E C=C + Nu$^{\ominus}$ $\longrightarrow$ E-C-C-Nu + X$^{\ominus}$

3+ $\quad$ X-C-C=C + Nu$^{\ominus}$ $\longrightarrow$ C=C-C-Nu + X$^{\ominus}$

Scheme 5.2 Some types of half-reactions

As useful as half-reactions are during the initial generalized phase of planning a synthesis, to advance further one needs a catalogue of real reactions that correspond to a given half-reaction. In order to organize the vast catalogue of reactions that correspond, for instance, to the reaction of $\mathbf{d}^1$-synthons with electrophiles, one may sort the type 1− half-reactions by the oxidation state at C-1 of the $\mathbf{d}^1$-synthon. One may further subclassify those by the kind of backbone of the $\mathbf{d}^1$-synthon, as adumbrated in Scheme 5.3.

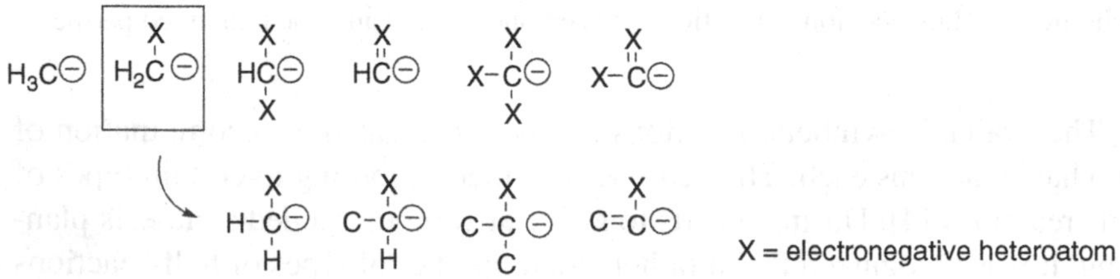

$H_3C^{\ominus}$ $\boxed{H_2C^{\ominus}}$ $HC^{\ominus}$ $HC^{\ominus}$ $X\text{-}C^{\ominus}$ $X\text{-}C^{\ominus}$

$H\text{-}C^{\ominus}$ $C\text{-}C^{\ominus}$ $C\text{-}C^{\ominus}$ $C=C^{\ominus}$ $\quad$ X = electronegative heteroatom

Scheme 5.3 Subgroups of $\mathbf{d}^1$-synthons

When desired, one may regard the oxidation state at C-2 of the $\mathbf{d^1}$-synthon, leading to a matrix of conceivable (desirable) reagents (Scheme 5.4).

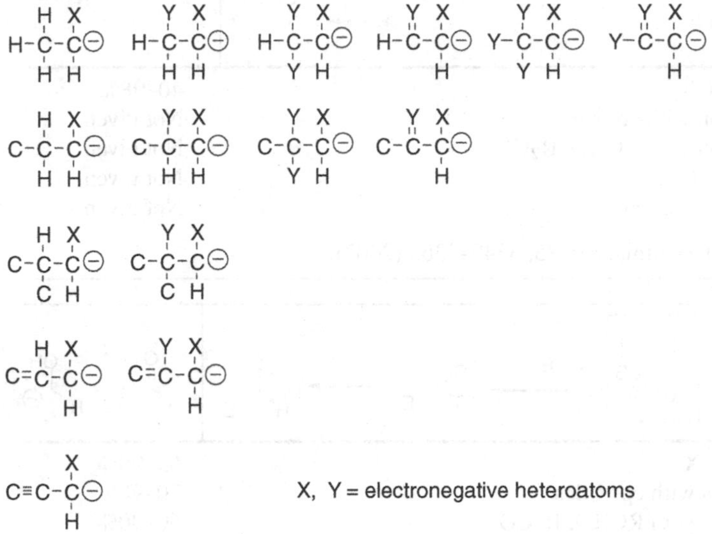

$$\begin{array}{cccccc}
\underset{\underset{H}{|}}{\overset{\overset{H}{|}}{H-C-}}\underset{H}{\overset{X}{\underset{|}{C}\ominus}} &
\underset{\underset{H}{|}}{\overset{\overset{Y}{|}}{H-C-}}\underset{H}{\overset{X}{\underset{|}{C}\ominus}} &
\underset{\underset{Y}{|}}{\overset{\overset{Y}{|}}{H-C-}}\underset{H}{\overset{X}{\underset{|}{C}\ominus}} &
\underset{H}{\overset{\overset{Y}{|}}{H-C-}}\overset{X}{\underset{|}{C}\ominus} &
\underset{\underset{Y}{|}}{\overset{\overset{Y}{|}}{Y-C-}}\underset{H}{\overset{X}{\underset{|}{C}\ominus}} &
\underset{H}{\overset{\overset{Y}{|}}{Y-C-}}\overset{X}{\underset{|}{C}\ominus}
\end{array}$$

H X   Y X   Y X   Y X   Y X   Y X
H-C-C⊖ H-C-C⊖ H-C-C⊖ H-C-C⊖ Y-C-C⊖ Y-C-C⊖
H H   H H   Y H   H   Y H   H

H X   Y X   Y X   Y X
C-C-C⊖ C-C-C⊖ C-C-C⊖ C-C-C⊖
H H   H H   Y H   H

H X   Y X
C-C-C⊖ C-C-C⊖
C H   C H

H X   Y X
C=C-C⊖ C=C-C⊖
H   H

X
C≡C-C⊖   X, Y = electronegative heteroatoms
H

Scheme 5.4 Matrix of desirable $\mathbf{d^1}$-synthons

Following such guidelines it becomes possible to arrive at a list of skeletal bond-forming reactions that belong to a certain type of half-reaction. The purpose of doing this is to check whether one or several real reactions exist to effect a transformation indicated by a half-reaction. Entries on such a list should contain the type of half-reaction, the synthon-type, and a table of typical skeletal bond-forming reactions possible with this reagent. Examples of how such entries might look are given below. More or less comprehensive lists of $\mathbf{d^1}$-synthons can be found in references [7, 8, 9].

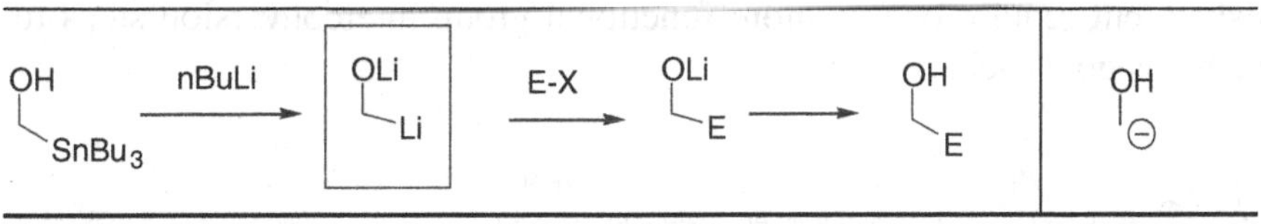

OH ──nBuLi──▶ OLi ──E-X──▶ OLi ──▶ OH   OH
  SnBu₃   Li   E   E   ⊖

Alkylation with R-X	40–98%
Hydroxyalkylation with epoxides	Not given
Hydroxyalkylation with RCHO, R₂CO	40–60%
Acylation with RCOX	Not given
1,4-Addtion to Enones, etc.	Not given

Seebach et al. Chem Ber 113, 1290–1303 (1980)

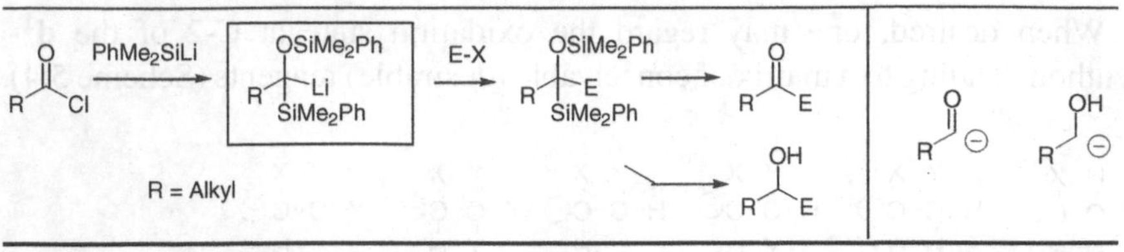

Alkylation with R-X	40–98%
Hydroxyalkylation with epoxides	Not given
Hydroxyalkylation with RCHO, R_2CO	Not given
Acylation with RCOX	Not given
1,4-Addtion to Enones, etc.	Not given

Fleming et al. Helv. Chim. Acta 85, 3349–3365 (2002)

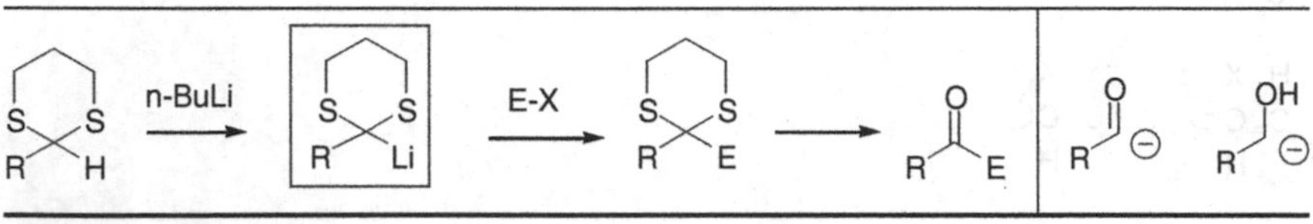

Alkylation with R-X	60–90%
Hydroxyalkylation with epoxides	70–90%
Hydroxyalkylation with RCHO, R_2CO	60–90%
Acylation with RCOX	problematic
1,4-Addtion to Enones, etc.	ca. 90%

Seebach et al. Synthesis 1969, 17–36
Brown et al. Chem. Comm. 1979, 100–101

During the detailed phase of planning a synthesis one would focus on those reagents, which would directly effect the desired transformation. Some desirable reagents, e.g., those of structure **38**, may not be available, because they contain a potential leaving group in the β-position to a negative charge. In such situations the *synthetic equivalents* of such reagents should be listed, i.e., reagents that ultimately allow the desired transformation to be accomplished, but require one or more functional group interconversion steps to reach this goal (Scheme 5.5).

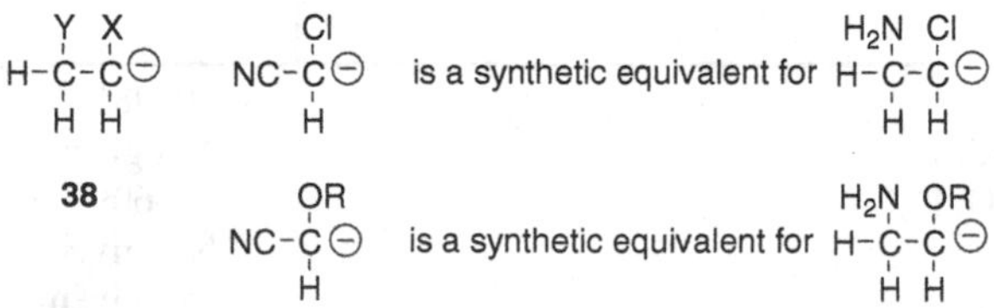

Scheme 5.5 Synthetic equivalents for synthons to which no analogous reagent is directly available

Some databases, such as WebReactions (http://www.webreactions.net/), REACCS (http://www.mdl.com/company/about/history.jsp) or MOS (http://www.accelrys.com/products/datasheets/chemdb_mos_a4.pdf) are organized in a similar fashion and allow one to search for real reactions that correspond to certain half-reactions.

The functional group oriented strategies, the skeleton oriented strategies, or the building block oriented strategies, discussed in the previous chapters, may concern only partial structures of a larger target molecule. Hence, there often remains the task of linking these partial structures together in order to reach the complete target. These linkage operations frequently require the use of bivalent conjunctive reagents [10], as discussed in Chap. 2.2.5. Therefore a list of such multiple coupling reagents [11] possessing various skeletons and functionalities will be handy at this stage of synthesis design. An example is given in Scheme 5.6, illustrating how ethyl acetoacetate may serve as a synthetic equivalent of the acetonyl-1,1-dianion synthon, because the former may be alkylated twice, followed by decarboxylation.

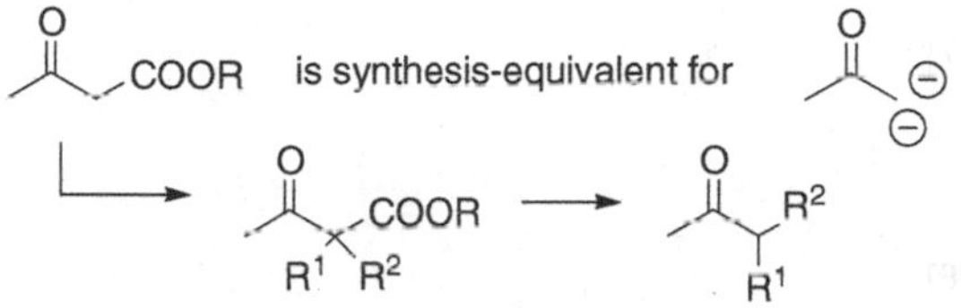

Scheme 5.6 Ethyl acetoacetate as synthetic equivalent for the acetonyl-1,1-dianion

Because bivalent conjunctive reagents are so useful for piecing together parts of a target structure, further examples (in addition to those given on p. 39) are shown in Scheme 5.7.

The value of using bivalent conjunctive reagents can be seen in a building block oriented synthesis of juvabione (**39**) [19] (Scheme 5.8).

Beyond *1,1*-bivalent conjunctive reagents, there are a variety of *1,2*- and *1,3*-bivalent conjunctive reagents [9, 11, 29]. Therefore, as a basis for planning syntheses, one not only needs a catalogue of differently functionalized synthons and corresponding real reagents, but also a similar catalogue of the various bivalent conjunctive reagents. Unfortunately, comprehensive catalogues of that sort do not yet exist.

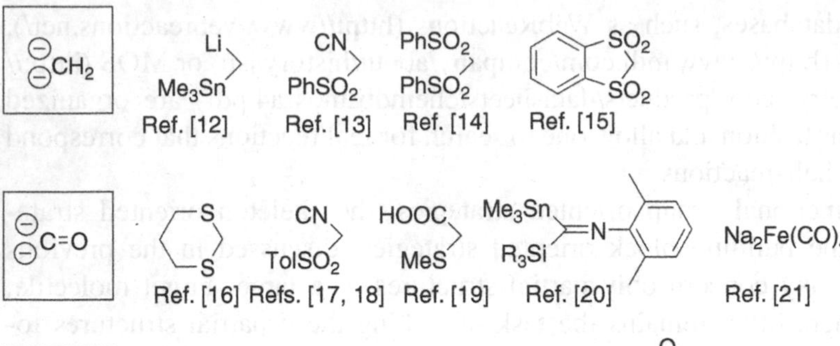

$\ominus\atop\ominus$CH₂ | Li / Me₃Sn | CN / PhSO₂ | PhSO₂ / PhSO₂ | [benzo-SO₂/SO₂ ring]

Ref. [12]   Ref. [13]   Ref. [14]   Ref. [15]

$\ominus\atop\ominus$C=O | [dithiane S–S] | CN / TolSO₂ | HOOC / MeS | Me₃Sn, R₃Si C=N–[aryl] | Na₂Fe(CO)₄

Ref. [16]   Refs. [17, 18]   Ref. [19]   Ref. [20]   Ref. [21]

$\oplus\atop\ominus$C=O | C=NᵗBu | Ph–C(=N)(–O)–Ph (N Ph oxazole) | Me₃Si–O–[benzo O₂S/SO₂]

Ref. [22]   Ref. [23]   Ref. [24]

$\oplus\atop\oplus$C=O | PhS, Cl–C=O | MeO–N, MeO–N–C=O

Ref. [25]   Ref. [26]

$\ominus\atop\ominus$C=CH₂ | SiMe₃ / H₂C–CH₂ / PhSO₂

Ref. [27]

$\oplus\atop\ominus$C=CH₂ | Ar–S(=O) / HC=CH₂

Ref. [28]

Scheme 5.7 Examples for synthetic equivalents of 1,1-bivalent conjunctive synthons

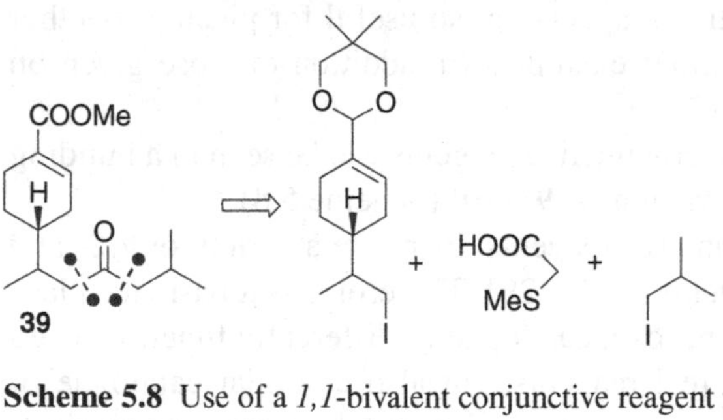

Scheme 5.8 Use of a *1,1*-bivalent conjunctive reagent in the synthesis of juvabione

Problems

5.1 When a nucleophilic and an electrophilic building block are to be linked in a synthesis by a methylene group (e. g., as a consequence of a building-block approach), a conjunctive reagent such as the one in Scheme 5.9 might be handy. What reagents and reactions could be used to effect such a coupling?

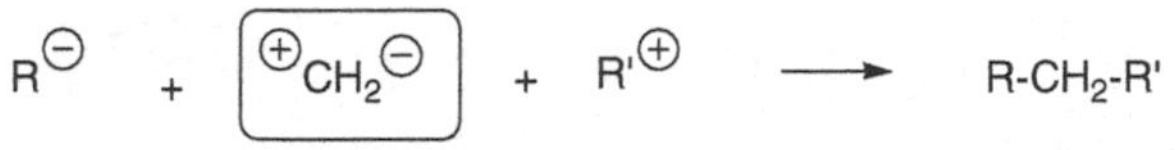

$$R^{\ominus} \quad + \quad \boxed{^{\oplus}CH_2^{\ominus}} \quad + \quad R'^{\oplus} \quad \longrightarrow \quad R\text{-}CH_2\text{-}R'$$

Scheme 5.9 Hypothetical conjunctive reagent for coupling of a nucleophilic and an electrophilic component

References

1. J. B. Hendrickson, *J. Am. Chem. Soc.* **1975**, *97*, 5784–5800.
2. G. Moreau, *Nouv. J. Chimie* **1978**, *2*, 187–193.
3. J. B. Hendrickson, *J. Chem. Educ.* **1978**, *55*, 216–220.
4. J. B. Hendrickson, *J. Chem. Inf. Comput. Sci.* **1979**, *19*, 129–136.
5. J. B. Hendrickson, D. L. Grier, A. G. Toczko, *J. Am. Chem. Soc.* **1985**, *107*, 5228–5238.
6. J. B. Hendrickson, T. M. Miller, *J. Am. Chem. Soc.* **1991**, *113*, 902–910.
7. T. A. Hase (ed.), *Umpoled Synthons*, J. Wiley & Sons, New York, **1987**, pp. 217–317.
8. T. A. Hase, J. K. Koskimies, *Aldrichimica Acta* **1981**, *14*, 73–77.
9. T. A. Hase, J. K. Koskimies, *Aldrichimica Acta* **1982**, *15*, 35–41.
10. B. M. Trost, D. M. T. Chan, *J. Am. Chem. Soc.* **1979**, *101*, 6429–6432.
11. D. Seebach, P. Knochel, *Helv. Chim. Acta* **1984**, *67*, 261–283.
12. T. Sato, S. Ariura, *Angew. Chem., Int. Ed. Engl.* **1993**, *32*, 105–106. (*Angew. Chem.* **1993**, *105*, 129–130).
13. J. S. Yadav, P. S. Reddy, *Tetrahedron Lett.* **1984**, *25*, 4025–4028.
14. D. Ferroud, J. M. Gaudin, J. P. Genet, *Tetrahedron Lett.* **1986**, *27*, 845–846.
15. E. P. Kündig, A. F. Cunningham Jr., *Tetrahedron* **1988**, *44*, 6855–6860.
16. D. Seebach, *Synthesis* **1969**, 17–36.
17. O. Possel, A. M. van Leusen, *Tetrahedron Lett.* **1977**, *18*, 4229–4232.
18. J. S. Yadav, L. Chetia, *Org. Lett.* **2007**, *9*, 4587–4589.
19. B. M. Trost, Y. Tamaru, *J. Am. Chem. Soc.* **1977**, *99*, 3101–3113.
20. Y. Ito, T. Matsuura, M. Murakami, *J. Am. Chem. Soc.* **1983**, *109*, 7888–7890.
21. J. P. Collman, S. R. Winter, D. R. Clark, *J. Am. Chem. Soc.* **1972**, *94*, 1788–1789.
22. G. E. Niznik, W. H. Morrison III, H. M. Walborsky, *J. Org. Chem.* **1974**, *39*, 600–604.
23. H. H. Wasserman, R. W. DeSimone, W.-B. Ho, K. E. McCarthy, K. S. Prowse, A. P. Spada, *Tetrahedron Lett.* **1992**, *33*, 7207–7210.
24. B. M. Trost, P. Quayle, *J. Am. Chem. Soc.* **1984**, *106*, 2469–2471.

25. C. Cardellicchio, V. Fiandanese, G. Marchese, L. Ronzini, *Tetrahedron Lett.* **1985**, *26*, 3595–3598.
26. W. L. Whipple, H. J. Reich, *J. Org. Chem.* **1991**, *56*, 2911–2912.
27. T. P. Meagher, L. Yet, C.-N. Hsiao, H. Shechter, *J. Org. Chem.* **1998**, *63*, 4181–4192.
28. H. Takei, H. Sugimura, M. Miura, H. Okamura, *Chem. Lett.* **1980**, 1209–1212.
29. S. D. Rychnovsky, D. Fryszman, U. R. Khire, *Tetrahedron Lett.* **1999**, *40*, 41–44.

Chapter 6
Formation of Cyclic Structures

Abstract Rings in a target structure are to be made from acyclic precursors by intramolecular one-bond formation (ring closure reaction) or by two-bond formation in a cycloaddition reaction. Bicyclic and polycyclic target structures are approached in the same way, whereby two-bond disconnections or multi-bond disconnections in reaction cascades are preferred. Multi-bond disconnections may be advantageous, even when a surplus extra bond is generated in the forward synthesis.

A ring in a target structure can be formed by a ring closure reaction from open-chain precursors forming one ring bond. Two ring bonds may be formed in one stroke, when the ring is formed by a cycloaddition reaction [1]. Hence, when addressing the formation of rings in retrosynthesis, both one-bond disconnections and two-bond disconnections have to be evaluated. Regarding cyclopropanes, both alternatives appear to be well precedented. This is also true for cyclobutanes, for which both photo-[2 + 2]- and ketene-[2 + 2]-cycloadditions are well established. For the formation of carbocyclic cyclopentanes or cycloheptanes, cycloadditions claim only a minor role, because [3 + 2]- and [4 + 3]-cycloaddition reactions [2] are not yet fully developed. Thus, for cyclopentanes and cycloheptanes ring closing reactions, such as intramolecular enolate alkylation and Dieckmann cyclizations, dominate. For an overview of ring forming reactions, see Scheme 6.1.

For one-bond disconnections of a ring (planning ring closure reactions), one selects the cut according to the functional group presence or the presence of substituents (= branches). For two-bond disconnections one could envision stepwise formation of these bonds. This leads one to look back at the bivalent conjunctive reagents presented in Chap. 2.2.5 and 5 (cf. Schemes 5.7, 2.62, and 2.63). These reagents are well-suited for the formation of cyclic structures and are frequently used in this context.

The available methodology for generating six-membered rings is quite varied and thus provides several options during the planning of a synthesis.

R.W. Hoffmann, *Elements of Synthesis Planning*,
DOI 10.1007/978-3-540-79220-8_6, © Springer-Verlag Berlin Heidelberg 2009

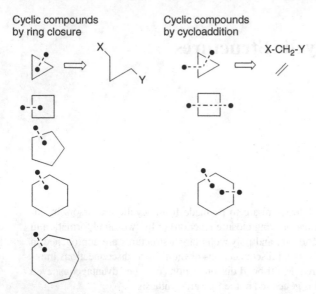

Scheme 6.1 Bond-sets for the construction of cyclic compounds

Principal possibilities [3] for the synthesis of cyclohexane **40** are illustrated in Scheme 6.2.

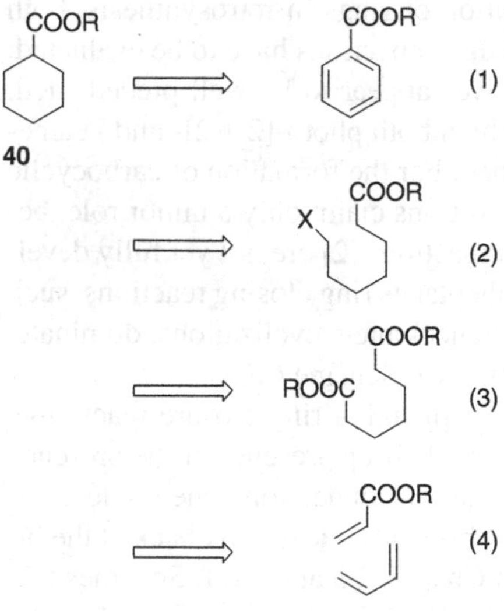

Scheme 6.2 Retrosynthesis of cyclohexane **40**

When a cyclohexane ring is present in the target, one should always determine whether or not a corresponding aromatic compound exists that could be reduced to the desired cyclohexane by either hydrogenation or by Birch reduction followed by hydrogenation (Scheme 6.2, case (1)). One should also consider the classical ring-forming reactions (Scheme 6.2, cases (2) and (3)). It is most attractive to form six-membered rings by [4 + 2]-cycloaddition reactions (Scheme 6.2, case (4)). Unfortunately, the strategically valuable [3 + 3]-cycloaddition methodology is essentially undeveloped [4, 5, 6, 7, 8, 9, 10, 11]. Hence, Diels-Alder cycloaddition is frequently the silver bullet. Regioselectivity issues arise when substituents are present on both the diene and the dienophile. The general rule is that a ψ-ortho or ψ-para arrangement of substituents is preferred over a ψ-meta arrangement in the resulting cyclohexene ring (Scheme 6.3).

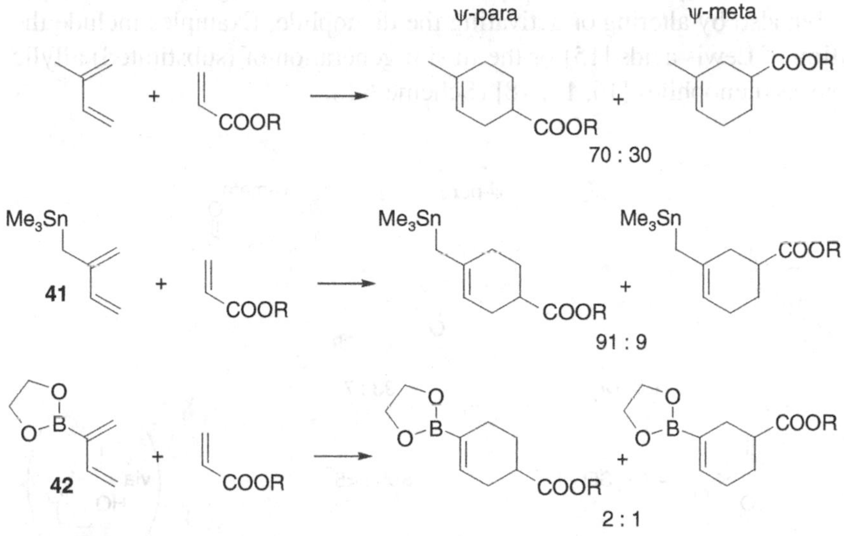

Scheme 6.3 Regioselectivity of Diels-Alder cycloadditions

The degree of regioselectivity is controlled in large part by the orbital coefficients in the HOMO of the diene and in the LUMO of the dienophile. One can exploit this aspect of the [4 + 2]-cycloadditions to enhance regioselectivity [12]. For instance, an auxiliary silyl or stannyl substituent (cf. compound **41**) may serve to improve an otherwise unsatisfactory regioselectivity during a Diels-Alder cycloaddition [13]. A boryl substituent, such as in diene **42**, defines a position in the Diels-Alder product at which other (alkyl or aryl) substituents can be introduced in follow-up steps; unfortunately, diene **42** confers poor regioselectivity.

Oxygen-substituted dienes generally afford highly regioselective Diels-Alder cycloadditions. This is a hallmark of the Danishefsky dienes [14] (Scheme 6.4).

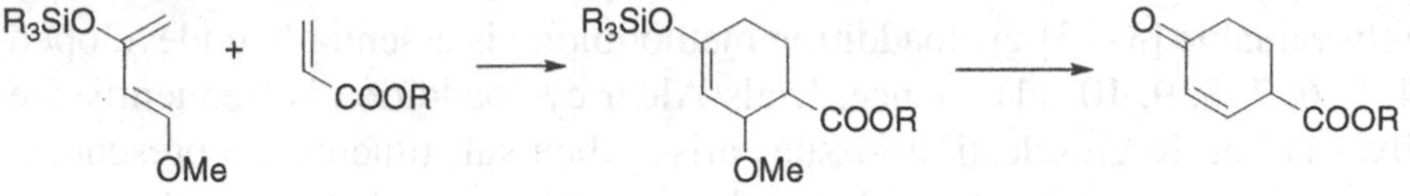

Scheme 6.4 High degree of regioselectivity through the use of Danishefsky dienes

Regioselectivity can be improved not only by modifying the diene component, but also by altering or activating the dienophile. Examples include the addition of Lewis acids [15] or the in situ generation of (substituted) allylic cations as dienophiles [16, 17, 18] (Scheme 6.5).

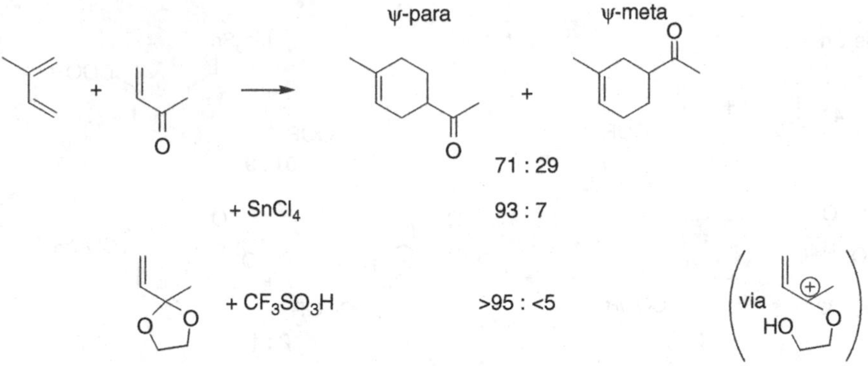

Scheme 6.5 Increase in regioselectivity by activation of the dienophile

In the end, the substituent patterns ψ-ortho and ψ-para can readily be obtained in Diels-Alder cycloadditions. When a ψ-meta arrangement of substituents is desired, one has to generate an umpolung by additional substituents. Umpolung at the diene [19, 20, 21, 22, 23] is illustrated in Scheme 6.6.

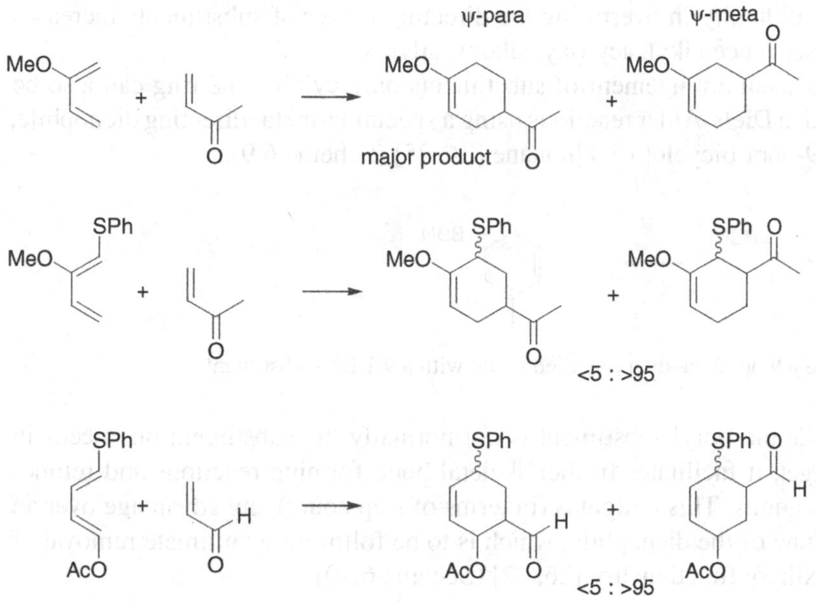

ψ-para   ψ-meta

major product

<5 : >95

<5 : >95

Scheme 6.6 Umpolung of regioselectivity in Diels-Alder cycloadditions

The arylthio group causes an umpolung that overrules the regiodirection of a methoxy or acetoxy group. Characteristic for an umpolung strategy, the arylthio substituent must be removed in a subsequent step, for instance by reduction with Raney-Ni.

In summary, for the generation of substituted six-membered carbocycles via the Diels-Alder cycoaddition there are "normal" dienes that result in a ψ-ortho and ψ-para arrangement of substituents in the product (Scheme 6.7):

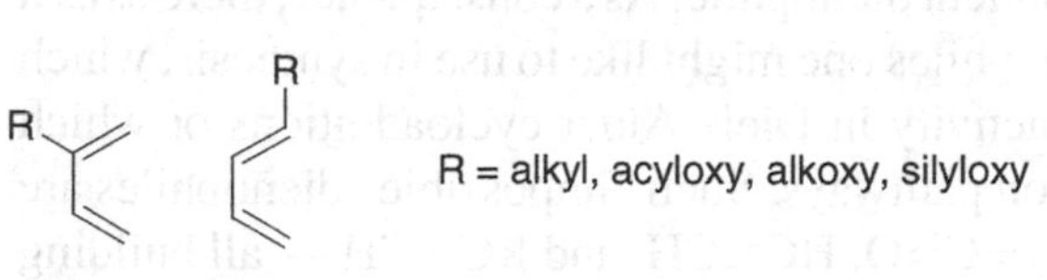

R = alkyl, acyloxy, alkoxy, silyloxy

Scheme 6.7 ψ-ortho and ψ-para directing "normal" dienes

and there are umpoled dienes, which allow the formation of cyclohexenes with a ψ-meta arrangement of the desired substituents (Scheme 6.8):

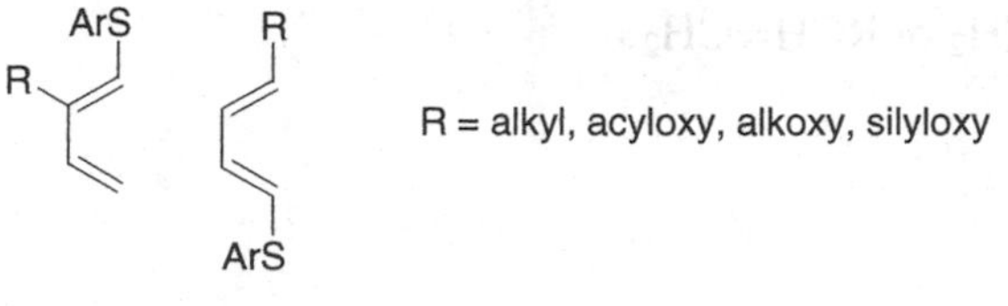

R = alkyl, acyloxy, alkoxy, silyloxy

Scheme 6.8 ψ-meta–directing dienes using umpolung by an arylthio group

The difficulty in overruling the directing power of substituents increases in the sequence alkyl, acyloxy, alkoxy, silyloxy.

A ψ-meta arrangement of substituents on a cylohexene ring can also be formed in Diels-Alder reactions using a special ψ-meta–directing dienophile, vinyl-9-bora-bicyclo[3.3.1]nonane [24, 25] (Scheme 6.9).

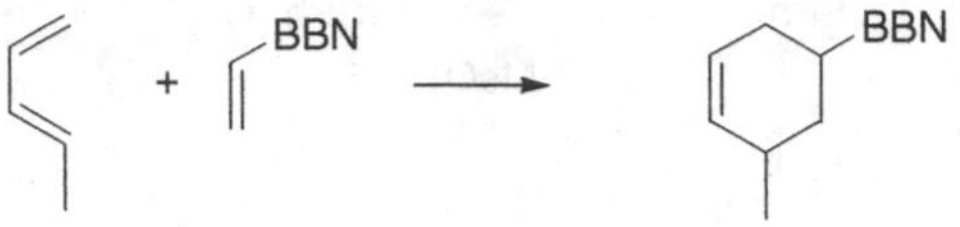

Scheme 6.9 ψ-meta–directing dienophile with a 9-BBN-substituent

While the boryl substituent is not normally the substituent one needs in the target, it facilitates further skeletal bond-forming reactions and refunctionalizations. This mitigates (in terms of step count) any advantage over an umpolung of the dienophile, which is to be followed by ultimate removal of the auxiliary functionality [26, 27] (Scheme 6.10).

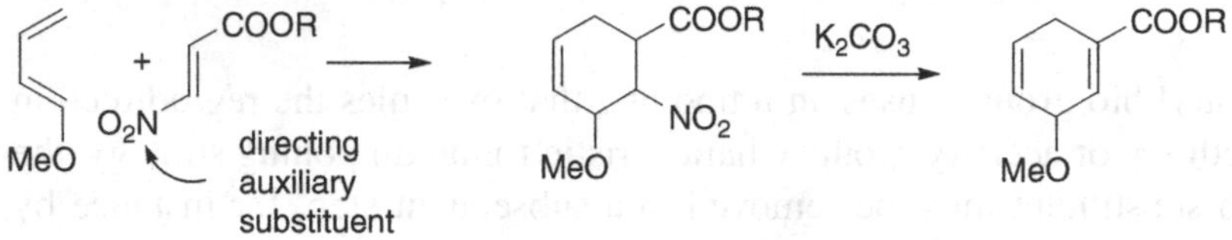

Scheme 6.10 Umpolung of a dienophile by a nitro group

Diels-Alder reactions (with normal electron demand) rely on an electron-rich diene and an electron-deficient dienophile. As a consequence, there exist a number of "impossible" dienophiles one might like to use in synthesis, which turn out to have too poor reactivity in Diels-Alder cycloadditions or which participate in alternate reaction pathways. Such "impossible" dienophiles are $CH_2=CH_2$, $RCH=CH_2$, $CH_2=C=O$, $HC\equiv CH$, and $RC\equiv CH$ — all building blocks that one really wishes to employ in the planning of a synthesis. Fortunately, a series of synthetic equivalents for these "impossible" dienophiles exists. They participate readily in Diels-Alder cycloadditions, though they require subsequent refunctionalization steps. The example [28] in Scheme 6.11 demonstrates that vinylsulfone $CH_2=CHSO_2Ph$ may serve as a synthetic equivalent for either $CH_2=CH_2$ or $RCH=CH_2$.

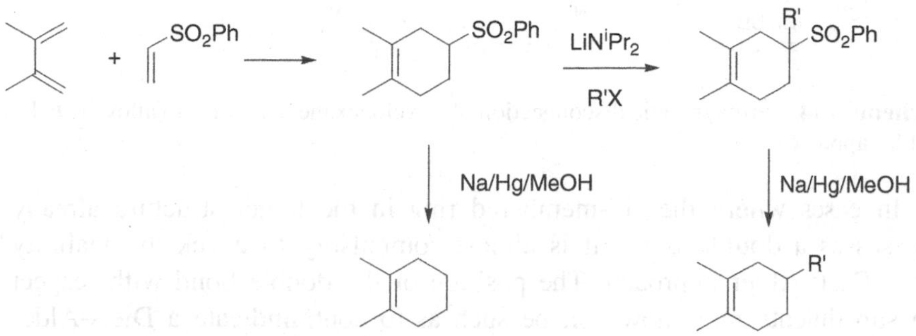

Scheme 6.11 PhSO$_2$CH=CH$_2$ as synthetic equivalent for CH$_2$=CH$_2$

In Scheme 6.12 are listed some typical synthetic equivalents for HC≡CH [29, 30, 31, 32].

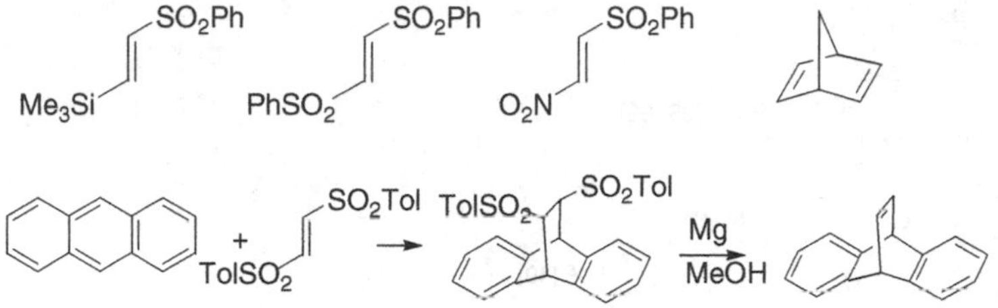

Scheme 6.12 Synthetic equivalents for HC≡CH

The scope of the Diels-Alder cycloaddition in synthesis planning is extended by the fact that both terminal and internal alkynes (RC≡CH, RC≡CR) enter into [4 + 2] cycloadditions with dienes in the presence of Co(0) catalysts. These cycloadditions result in cyclohexa-1,4-dienes arising from a formal Diels-Alder reaction [33, 34]. Finally, numerous synthetic equivalents for ketene to be used in [4 + 2] cycloadditions have been developed in the context of prostaglandin syntheses. Thus, a broad range of possibilities for achieving these cycloadditions is now available [35, 36, 37] (Scheme 6.13).

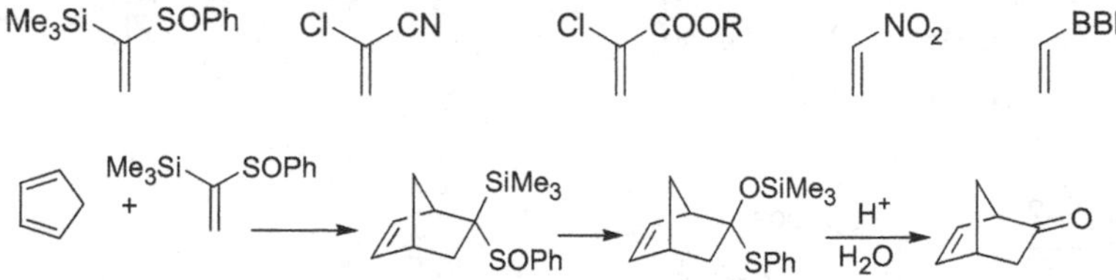

Scheme 6.13 Synthetic equivalents for H$_2$C=C=O

In order to check whether the Diels-Alder disconnection fits a cyclohexane moiety in a target structure, one starts by drawing a double bond in the ring (add DB). This bond should be placed in such a manner that the substituent pattern on the resulting cyclohexene ring could be reached by the reaction of normal dienes and dienophiles without resorting to umpoled variants or other synthetic equivalents. An example is given in Scheme 6.14.

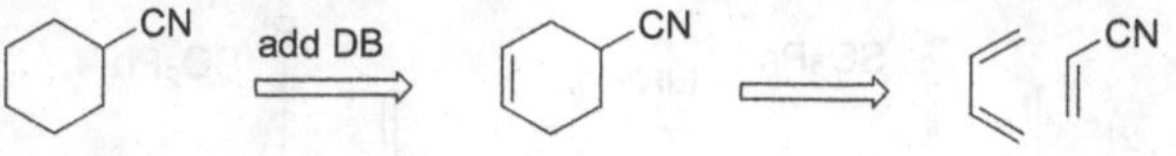

Scheme 6.14 Retrosynthetic disconnection of a cyclohexane derivative to allow a Diels-Alder approach

In cases where the six-membered ring in the target structure already possesses a double bond, it is almost compulsory to check the viability of a Diels-Alder approach. The position of the double bond with respect to substituents may, however, be such as to contraindicate a Diels-Alder

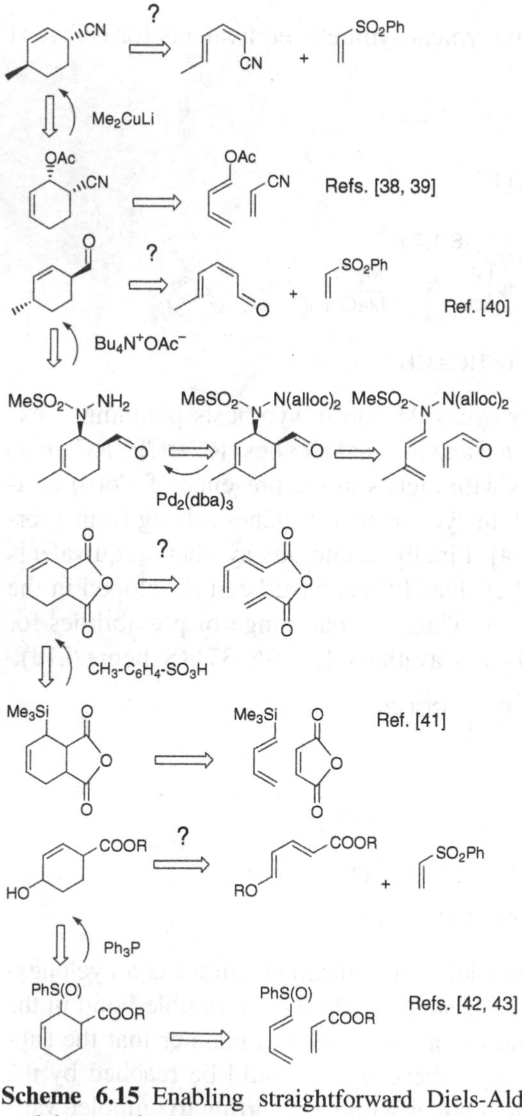

Scheme 6.15 Enabling straightforward Diels-Alder cycloadditions by shifting the cyclohexene double bond by one position

disconnection. In such cases it is good to know that reactions exist by which the double bond in the ring may be moved by one position (Scheme 6.15), hopefully resolving the problem.

In summary, there are a multitude of methods available for preparing six-membered rings with various substituent patterns. The same cannot be said for the synthesis of seven-membered rings. As a result, seven-membered rings are frequently accessed via *ring enlargement* [44] (Scheme 6.16), provided the required substituent pattern can be easily placed on the cyclohexane nucleus [45].

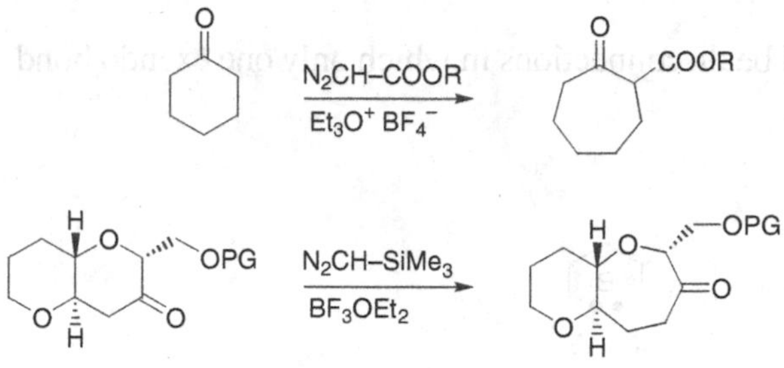

Scheme 6.16 Ring enlargement from six- to seven-membered rings

6.1 Anellated Bicycles and Anellated Polycycles

A bicyclic system consisting of two anellated rings can be viewed as a monocycle, which carries two substituents that happen to be closed to the other ring. In terms of retrosynthetic considerations, the four exendo bonds are important, i.e., those bonds that are endocyclic in one ring and exocyclic with respect to the other ring (Scheme 6.17).

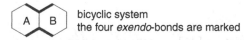

bicyclic system
the four *exendo*-bonds are marked

Scheme 6.17 Illustration of exendo bonds in an anellated bicyclic system

The formation of an exendo bond rapidly increases complexity during synthesis, as it creates a branch (a substituent) on one ring and at the same time prepares for ring closure of the second ring. Hence, the bond-set for anellated bicycles is chosen so as to form two exendo bonds. In this sense, the second ring is actually anellated to the first ring (Scheme 6.18).

Scheme 6.18 Bond-sets for the synthesis of anellated bicycles focusing on exendo bonds

Of lower ranking would be disconnections in which only one exendo bond is targeted (Scheme 6.19).

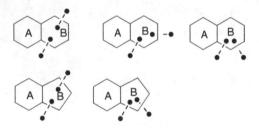

Scheme 6.19 Bond-sets for the synthesis of anellated bicycles featuring one exendo bond

Retrosynthetic analysis of anellated bicycles in this manner does not consider the endoendo bond, which is also called a fusion bond (Scheme 6.20).

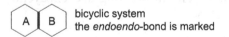

bicyclic system
the *endoendo*-bond is marked

Scheme 6.20 Illustration of the endoendo bond in an anellated bicycle

Since the classical Robinson annulation [46] numerous reaction schemes have been developed that allow the anellation of five- and of six-membered rings [47, 48, 49]. The following compilation (Scheme 6.21) illustrates that certain anellation procedures require and/or generate certain functionalities in either the original or the anellated ring.

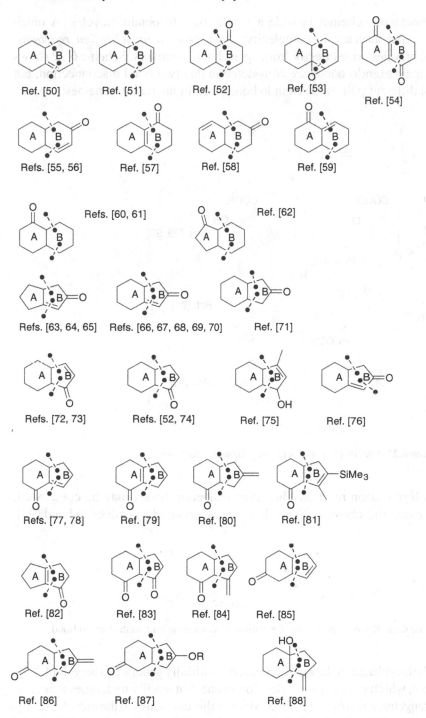

Scheme 6.21 Bond-sets for anellation schemes to obtain functionalized bicycles

Annelation schemes provide a viable route to obtain bicycles. A much more rapid increase in complexity is achieved in *bicyclization reactions*. Here, bicycles arise directly from open chain precursors. Scheme 6.22 shows that again exendo bonds are considered in this two-bond disconnection, but it is a different pair of the exendo bonds than in anellation schemes:

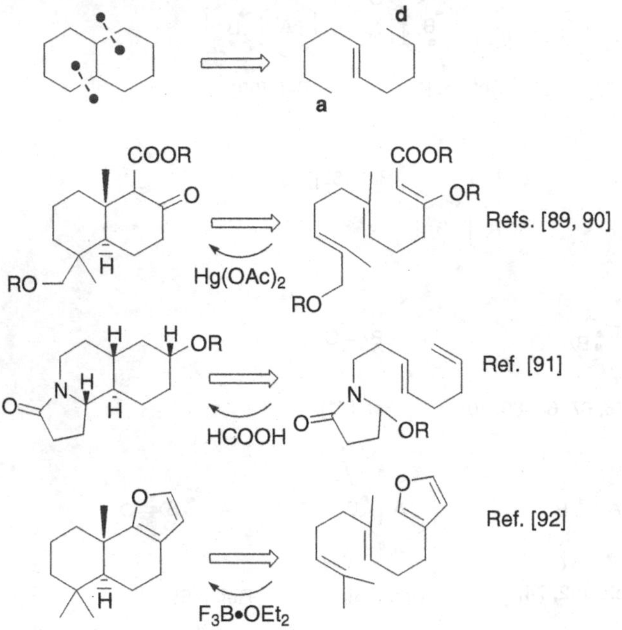

Scheme 6.22 Electrophile-induced bicyclization reactions

In biyclization reactions, however, endoendo bonds may be cut as well, cf. the case in Scheme 6.23 involving an exendo *and* an endoendo bond [93].

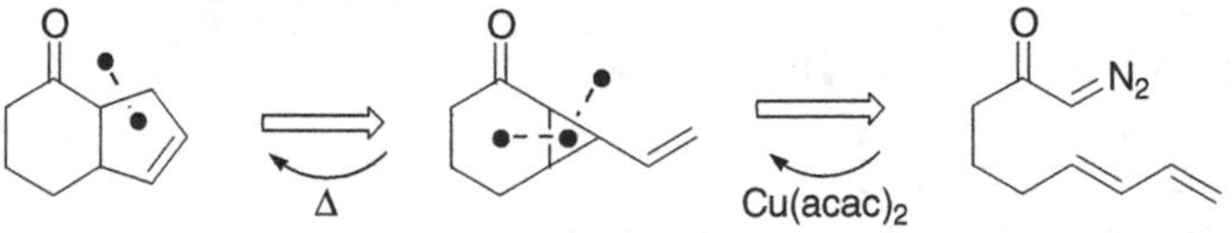

Scheme 6.23 Bicyclization with formation of an exendo and an endoendo bond

The bicyclization shown in Scheme 6.23 initially generates a bicyclo[4.1.0]-system, which contains a vinylcyclopropane that readily undergoes a thermal rearrangement to a cyclopentene, which in this case leads to the bicyclo[4.3.0]-system.

A highly favored reaction to effect bicyclizations is the intramolecular Diels-Alder cycloaddition [94] (Scheme 6.24), that forms an endoendo bond.

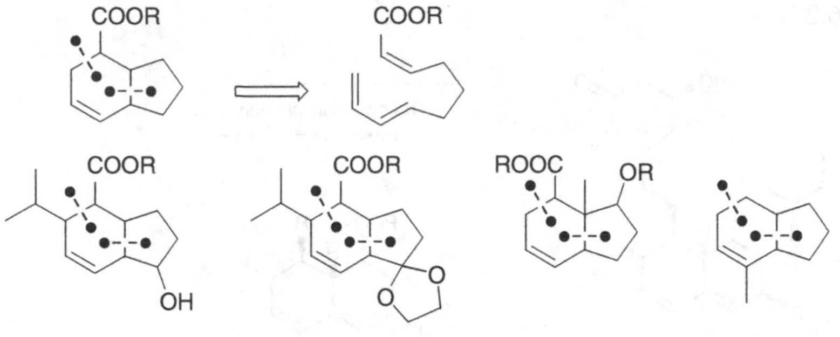

Scheme 6.24 Bicyclizations via intramolecular Diels-Alder cycloaddition

When the target bicycle lacks a suitably placed double bond in a six-membered ring, retrosynthesis starts with FGA (= add double bond) [95, 96, 97, 98]. In bicyclization reactions forming an endoendo bond, the stereochemistry of the reaction has to be monitored (formation of a *cis* or of a *trans* fusion of the two rings). With anellated systems of a five- to a six-membered ring the cis juncture is thermodynamically favored. This renders it possible to reach a cis fused bicyclo[4.3.0]nonane system by an epimerization process following the bicyclization reaction [99] (Scheme 6.25).

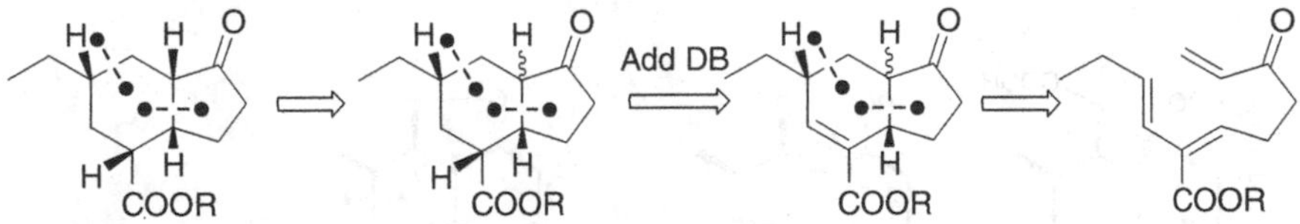

Scheme 6.25 Bicyclization followed by adjustment of the relative configuration (epimerization) at the endoendo bond

A bicyclization strategy allows one to create a complex pattern of substituents and functional groups by the methods of acyclic synthesis. This is then transformed in the final step (Scheme 6.26) into the target with a rapid increase in complexity [100].

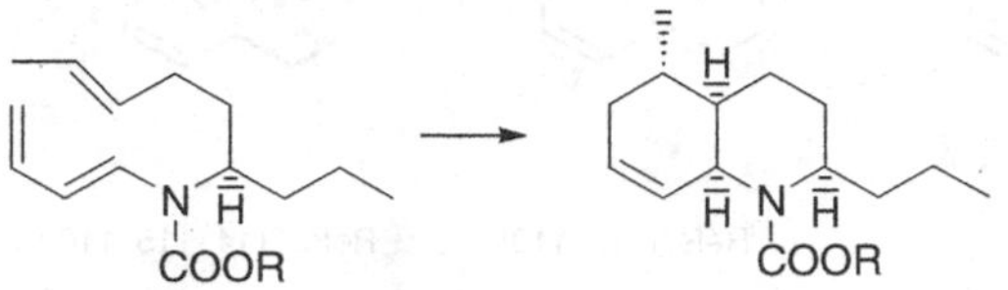

Scheme 6.26 Rapid increase in complexity by bicyclization of an open-chain precursor

To judge the relative merits of an anellation versus a bicyclization strategy, it is instructive to look at the many syntheses [101] of compactin and mevinolin, which represented a popular synthesis target in the early 1980s (Scheme 6.27).

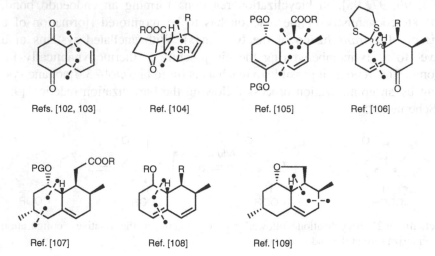

Anellations:

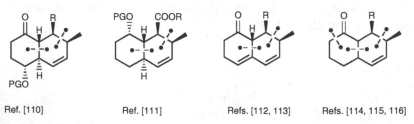

Scheme 6.27 Anellation versus bicyclization during syntheses of compactin and mevinolin

When looking at polycyclic target structures like those of the tetracycline or anthracycline antibiotics, the most efficient syntheses rely on bicyclization approaches. They are especially impressive when they have been executed as cascade reactions forming more than two bonds at a time. In this context, the reader is referred to Muxfeldt's 1965 synthesis of tetracycline (Scheme 6.28) [117], the first example of this groundbreaking strategy.

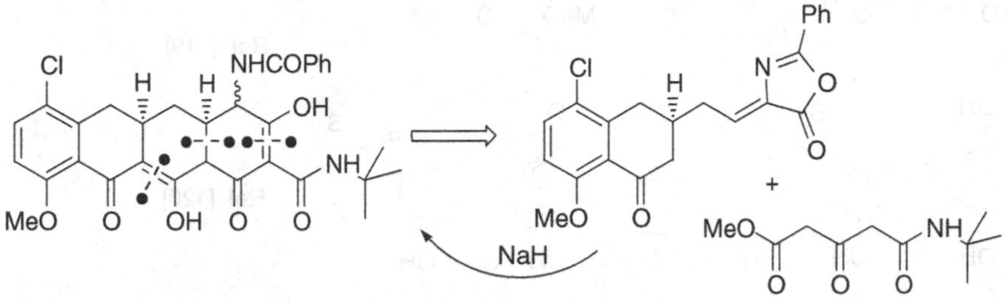

Scheme 6.28 Rapid increase in complexity by bicylization during Muxfeldt's synthesis of tetracycline

The elegance of this approach becomes obvious when compared to the sequential anellations used by Woodward in his tetracycline synthesis [118] (Scheme 6.29).

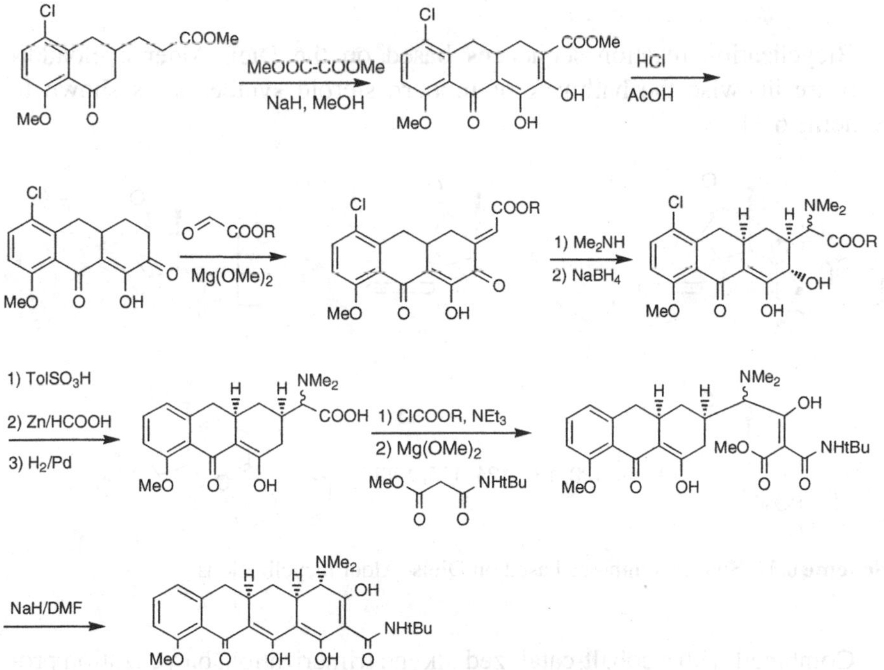

Scheme 6.29 Woodward's construction of the tetracycline skeleton by sequential anellation reactions

The following examples further illustrate the versatility of bicyclization strategies (Scheme 6.30).

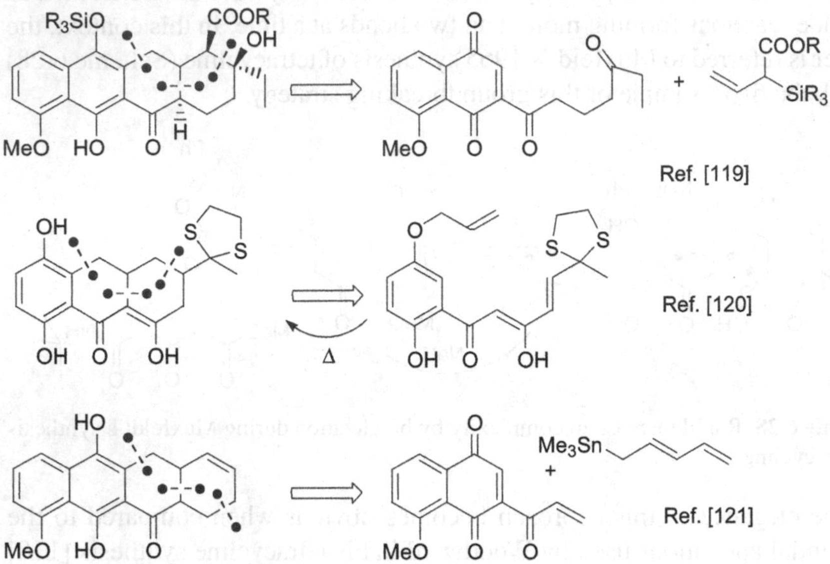

Ref. [119]

Ref. [120]

Ref. [121]

Scheme 6.30 Rapid increase in complexity by bicylization reactions

Bicyclization reaction sequences based on the Diels-Alder cycloaddition are likewise the hallmark of modern steroid syntheses, as shown in Scheme 6.31.

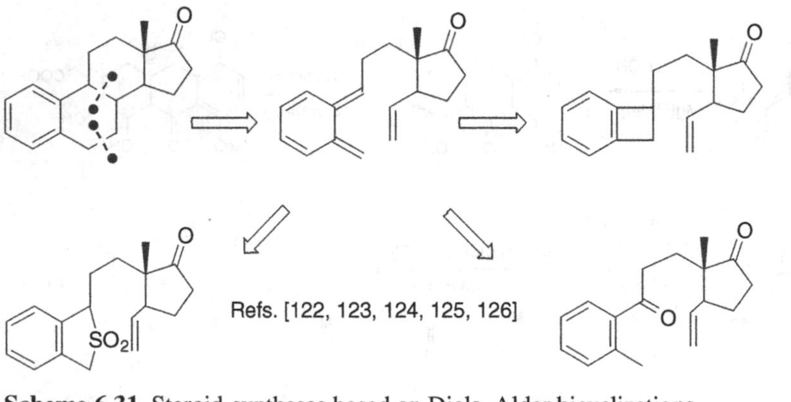

Refs. [122, 123, 124, 125, 126]

Scheme 6.31 Steroid-syntheses based on Diels–Alder bicyclizations

Combined with a cobalt-catalyzed alkyne trimerization, bicyclization provides spectacular inroads to the skeleton of estrone (Scheme 6.32) [127].

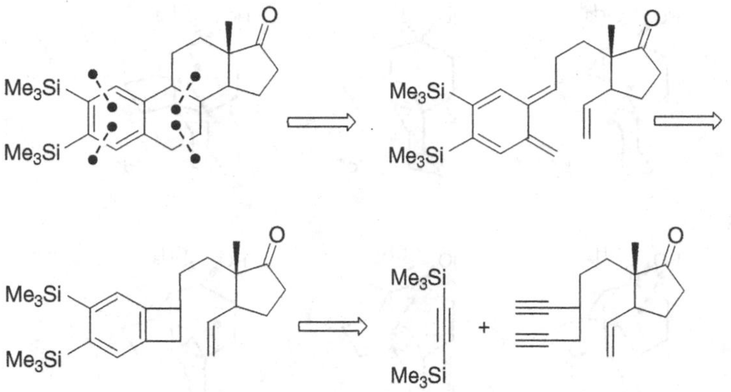

Scheme 6.32 Bicyclization reaction as a key step in a rapid synthesis of the estrone skeleton

When considering the synthesis of anellated bi- and polycycles, the rapid increase in complexity associated with bicyclization reactions suggests that an evaluation of this possibility should be considered first. Therefore, determine whether the target contains a centrally located six-membered ring that can be constructed using an intramolecular Diels–Alder cycloaddition.

6.2 Bridged Bi- and Polycycles

Bridged polycycles, such as compound **43** (Scheme 6.33), have complex molecular skeletons. In turn, retrosynthetic approaches are not immediately obvious. A more systematic approach, such as the one developed by the Corey group in the 1970s, is required [128]. Key elements of this approach are discussed with respect to compound **43** in Scheme 6.34.

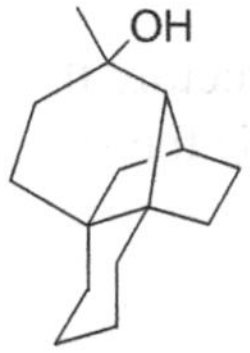

Scheme 6.33 Example of a bridged polycyclic ring system

In order to initiate the discussion, several skeletal bonds of **43** have been arbitrarily picked in Scheme 6.34. We would like to know which of these cuts results in the greatest retrosynthetic simplification of the target.

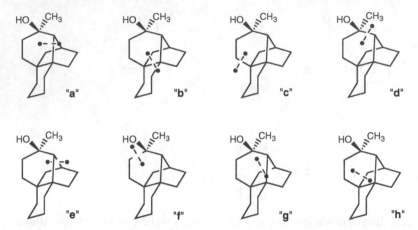

Scheme 6.34 Trial and error approach to identify an optimal retrosynthetic cut in a bridged polycycle

Retrosynthetic simplification is reached by reducing the number of bridges in the target structure. While every one of the proposed cuts achieves this goal, some can readily be given lower priority. Cuts "a" and "b" affect endoendo bonds. Their cleavage generates precursor molecules with bridged middle-sized rings, e.g., a bridged ten-membered carbocycle on cut "b." Such cuts are deemed unattractive, because the precursor synthesis (aside from all the remaining bridges) is considered to be prohibitively complicated. Hence, cuts at endoendo bonds that generate medium-sized ring precursors are rejected. Cuts at exendo bonds such as "c," "d," "f," "g," and "h" generate rings with pendant side-chains, which in the cases of "c," "d," and "f" contain a stereogenic center requiring further attention. These considerations can be summarizeded by postulating that an optimal cut should:

- reduce the number of bridges;
- avoid medium-sized rings in the intermediates;
- minimize the number of pendant chains.

In order to meet these goals one should identify in the target structure the most highly bridged ring [128]. In the case of compound **43**, this is the ring marked "i" in Scheme 6.35.

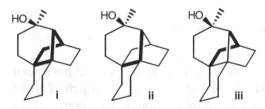

Scheme 6.35 The most highly bridged ring in compound **43**

In the present example, every bond in this ring is an exendo bond, the breaking of which will reduce the number of rings in the targeted structure. The bonds marked in "ii" are in addition endoendo bonds, so-called *core bonds*, the cutting of which would generate a medium-sized ring. This reduces further considerations to the remaining *strategic bonds* marked in "iii" (= i minus ii). Of these, the cut of a "zero-atom-bridge" (bond "e" in Scheme 6.34), leads to a simplified ring system without generating pendant chains. The cutting of a bridge with a length of one or more atoms ("g" or "h" in Scheme 6.34) leads to pendant chains, a result which is considered less desirable.

In summary, first try to identify the most highly bridged ring in the target structure. If one or more of the bonds in this ring is a core bond, it is rejected. Of the remaining bonds in this ring, identify the exendo bonds, because these represent strategic bonds. Only when several possibilities remain at this stage does one check whether any of these bonds are contained in a ring with stereogenic centers. As the formation of these rings would be accompanied by stereochemical issues, these cuts are given a lower ranking. In the case of **43**, cut "e" (Scheme 6.34) would therefore be given priority over cuts "g" and "h" (based on the number of pendant chains).

Example **43** given above is related to the synthesis of longifolene (**44**), which in the late 1970s was a challenging target for synthesis. It should be easy to apply the above rules to this target (Scheme 6.36).

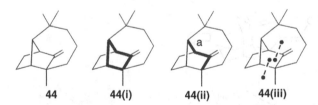

44   **44(i)**   **44(ii)**   **44(iii)**

Scheme 6.36 Search for strategic bonds for a synthesis of longifolene

The most highly bridged ring is readily identified (i). After exclusion of the core bonds, there remain the strategic bonds (ii). Of these, bond "a" as a zero-atom-bridge should be given highest priority. A review [129] of seven successful longifolene syntheses shows that, in three cases, the formation of bond "a" was the key step. The analysis introduced above pertains to one-bond disconnections. When one considers two-bond disconnections, i.e., bi-cyclization strategies, the eye is drawn to disconnection (iii) in Scheme 6.36. The advantages and disadvantages of this approach have been exhaustively discussed in reference [129]. Note that on two-bond disconnection (bicyclization), the cut of core bonds is frequently highly advantageous!

A bridged polycyclic target of more recent vintage is FR901483 (**45**), shown in Scheme 6.37. The most highly bridged ring and the strategic bonds are readily identified, as in **46**. In order to minimize the number of pendant chains resulting from a retrosynthetic cut, exendo bonds "a" or "c" appear to be more favorable than bond "b."

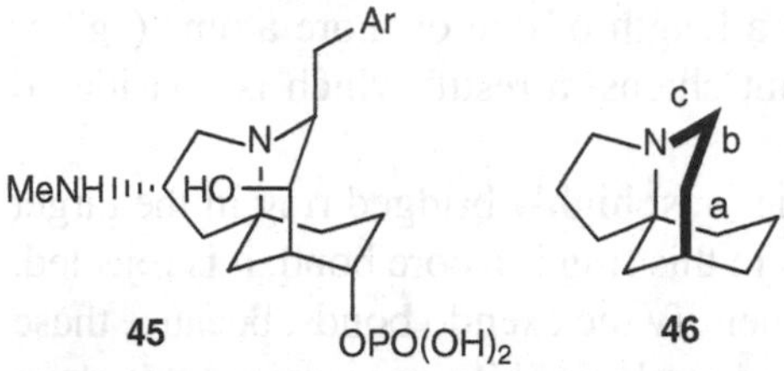

Scheme 6.37 Search for strategic bonds for a synthesis of FR901483

It is noteworthy that five [130, 131, 132, 133, 134] out of six syntheses of FR901483 used the formation of bond "a" as a key step to generate the complex molecular skeleton. The only "deviating" synthesis [135] utilized a bicyclization cascade sequence summarized in Scheme 6.38.

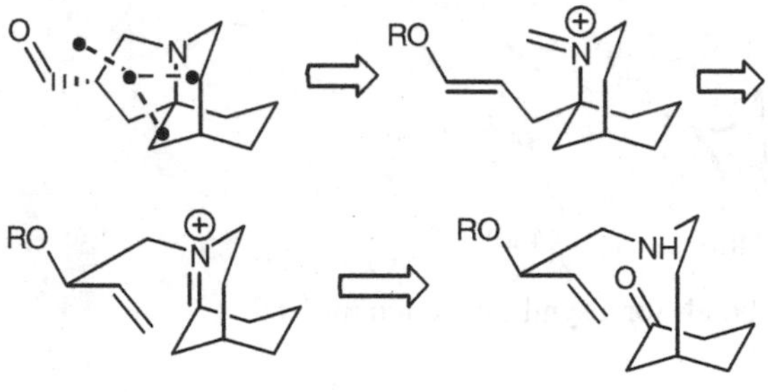

Scheme 6.38 Bicyclization approach to FR901483

The bridged polycyclic target sarain (**47**) (Scheme 6.39) provides an even greater challenge for synthesis. The initial approaches concentrated on inroads to the tricyclic core (**47i**) of sarain.

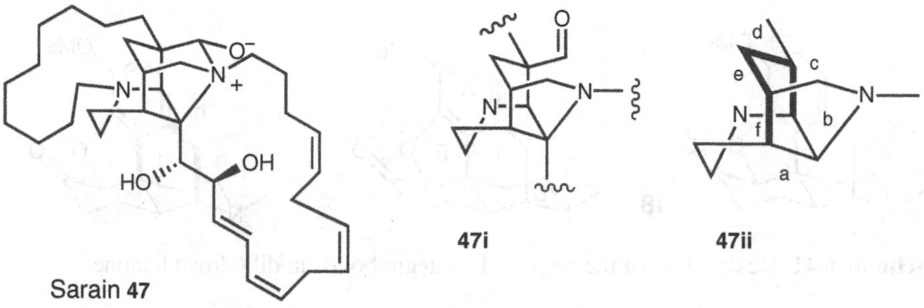

Sarain 47

Scheme 6.39 Search for strategic bonds for a synthesis of the sarain core

The identification of the most highly bridged ring reveals the strategic bonds shown in **47ii**. A cut at bond (d) would generate two pendant chains, that at bond (f) an anellated 6/7-ring system. The greatest simplification is realized by a retrosynthetic cut at bond (c), revealing an anellated 6/5-ring system as suitable precursor. Again, in most syntheses aiming at sarain or the sarain core, formation of bond (c) was successfully employed (Scheme 6.40).

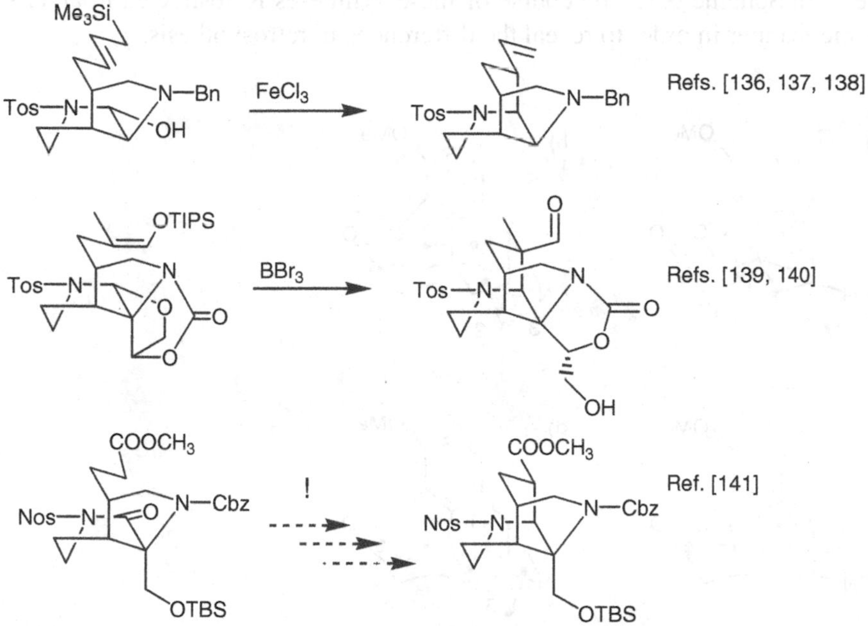

Scheme 6.40 Routes used to access the polycyclic core of sarain

Deeper insight into the possibilities for constructing bridged polycyclic target structures can be gained by studying the syntheses of morphine [142] and its key precursor dihydrocodeinone (**48**), shown in Scheme 6.41. In **48**, the rings A, E, and C are anellated, whereas rings B and D are bridged.

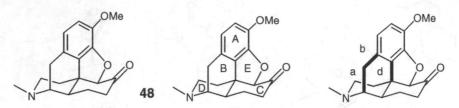

Scheme 6.41 Designation of the rings and strategic bonds in dihydrocodeinone

The most highly bridged ring is B, containing the strategic bonds (a) and (b). Bond (d) is a core bond between rings B and E. When one considers anellating ring E at the end of the synthetic sequence, then bond (d) loses the status of a core bond and becomes a strategic bond as well.

It is interesting to note the extent to which such considerations are reflected in recent syntheses of dihydrocodeinone and its congeners. It is instructive to read and reflect upon these syntheses in the original papers. Here, in Scheme 6.42, the course of these syntheses is abstracted in an extreme manner in order to reveal the differences in retrosynthesis.

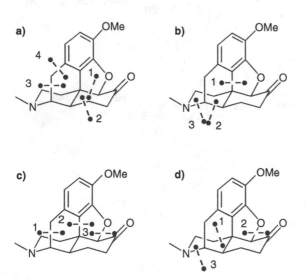

Scheme 6.42 Bond-sets of recent syntheses of dihydrocodeinone, (a) by D. A. Evans [143], (b) by K. A. Parker [144, 145] and B. M. Trost [146] (codeine), (c) by L. E. Overman [147, 148], and (d) by J. Mulzer [149]

Evans forms two of these strategic bonds late in his synthesis (Scheme 6.42a). Overman forms these bonds early. The syntheses of Parker and that of Trost form the bridged bicyclic nucleus rather late, without using the strategic bonds identified before. This holds in particular for the synthesis by Mulzer (Scheme 6.42d). This emphasizes a strategy for constructing bridged polycyclic systems that differs completely from Corey's approach to identifying strategic bonds. In this alternate approach, the most highly bridged ring is made first (or early). Then, the bridging and anellated rings are attached sequentially, as in a crocheting endeavor. This alternate approach features prominently in syntheses of quadrone (**49**), which was a popular target for synthesis in the 1980s (Scheme 6.43).

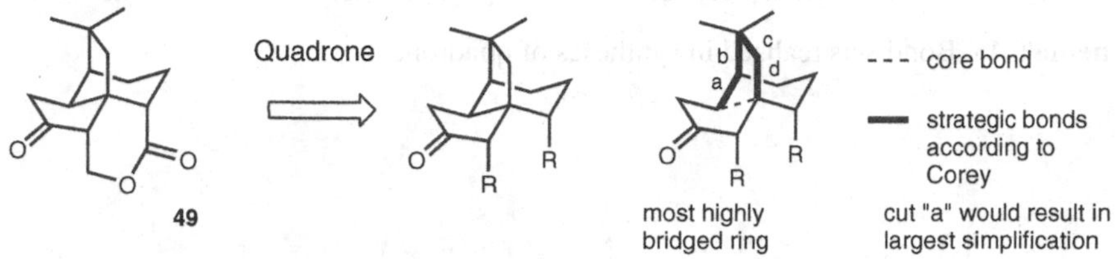

Scheme 6.43 Search for strategic bonds for a synthesis of quadrone

Of the four rings in quadrone, the lactone ring is closed last in all syntheses. Thus, one concentrates on the construction of the bridged tricyclic core of quadrone. Following Corey's analysis, the most highly substituted ring and the strategic bonds are easily identified. A retrosynthetic cut at bond (a) would result in the largest simplification. A cut at bond (c) would generate two pendant chains. Surprisingly, though, there is only one synthesis of quadrone, which closes strategic bond (a) [150] (Scheme 6.44), albeit in an impressive cascade reaction.

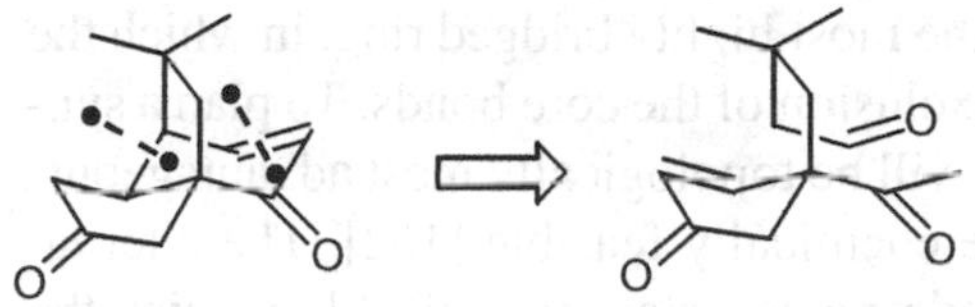

Scheme 6.44 Bicyclization in a synthesis of the quadrone skeleton

By contrast, most of the other syntheses of quadrone start from the most highly bridged ring, to which the other rings are then added. These syntheses, abstracted in Scheme 6.45, employ a broad variety of methods, as well as a cascade bicyclization scheme illustrated in Scheme 6.46 [151].

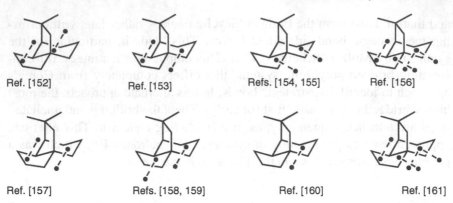

Ref. [152]   Ref. [153]   Refs. [154, 155]   Ref. [156]

Ref. [157]   Refs. [158, 159]   Ref. [160]   Ref. [161]

Scheme 6.45 Bond-sets realized in syntheses of quadrone

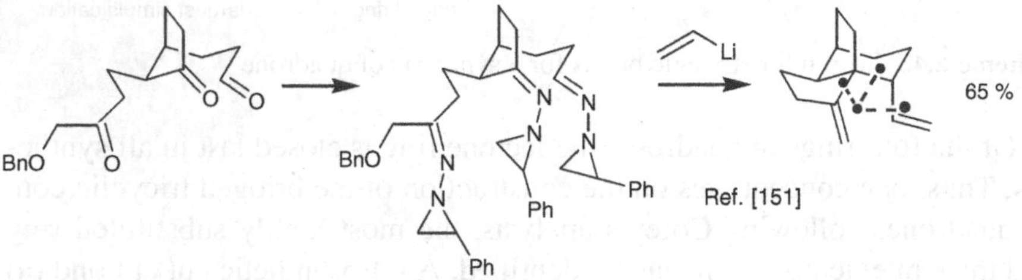

65 %

Ref. [151]

Scheme 6.46 Cascade bicyclization reaction yielding the quadrone skeleton

In summary, when faced with bridged polycyclic target structure, consider one-bond disconnections first, following the retrosynthetic analysis scheme by Corey. This analysis focuses on the most highly bridged ring, in which the strategic bonds are identified after exclusion of the core bonds. To plan a synthesis that forms this strategic bond will be topologically most advantageous. However, it may not necessarily be chemically feasible [162]. The alternative is to use the most highly bridged ring as a starting scaffold to which the other rings are sequentially added. Especially when the target structure contains six-membered rings, consider two-bond disconnections as well. These allow for intramolecular Diels-Alder cycloadditions, by which critical core bonds could be formed!

The possibilities of using two-bond disconnections (i.e., the Diels-Alder transform) are easy to appreciate with respect to patchouli alcohol (**50**), a bridged tricyclic target. It is a historically amusing compound, as its initial (wrong) structure was "proven" by synthesis [163] (a proof that later had to be revised)! [164] The correct structure of **50** possesses three bridged six-membered rings. A Diels-Alder transform can be initiated by: "Add DB" = addition of a double bond into one of the six-membered rings. The three possibile ways [165] to do this are illustrated in Scheme 6.47. This then sets the stage for an ensuing two-bond disconnection, providing the precursor for an intramolecular Diels-Alder cycloaddition [165].

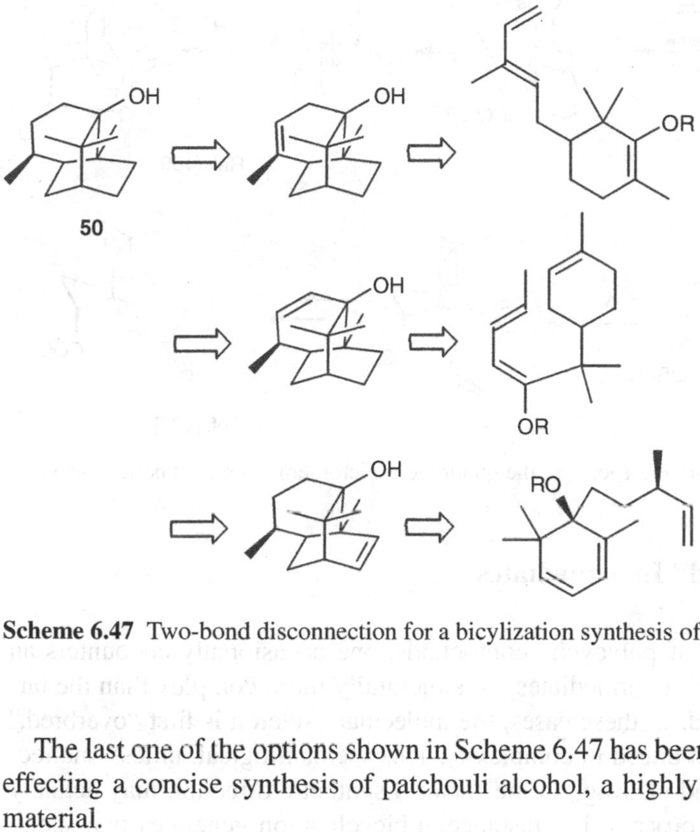

Scheme 6.47 Two-bond disconnection for a bicylization synthesis of patchouli alcohol

The last one of the options shown in Scheme 6.47 has been realized [166], effecting a concise synthesis of patchouli alcohol, a highly valued fragrant material.

The discussion in this text has presented rules by which it becomes possible to plan syntheses in a reasonable manner. This textbook approach to retrosynthesis stands in contrast to many surprising artistic solutions to synthesis problems. Such solutions may arise by imaginative incorporation of skeletal rearrangements into the planning of syntheses [167]. This holds particularly for the synthesis of bridged polycyclic structures. Scheme 6.48 illustrates some examples relating to the synthesis of the quadrone skeleton.

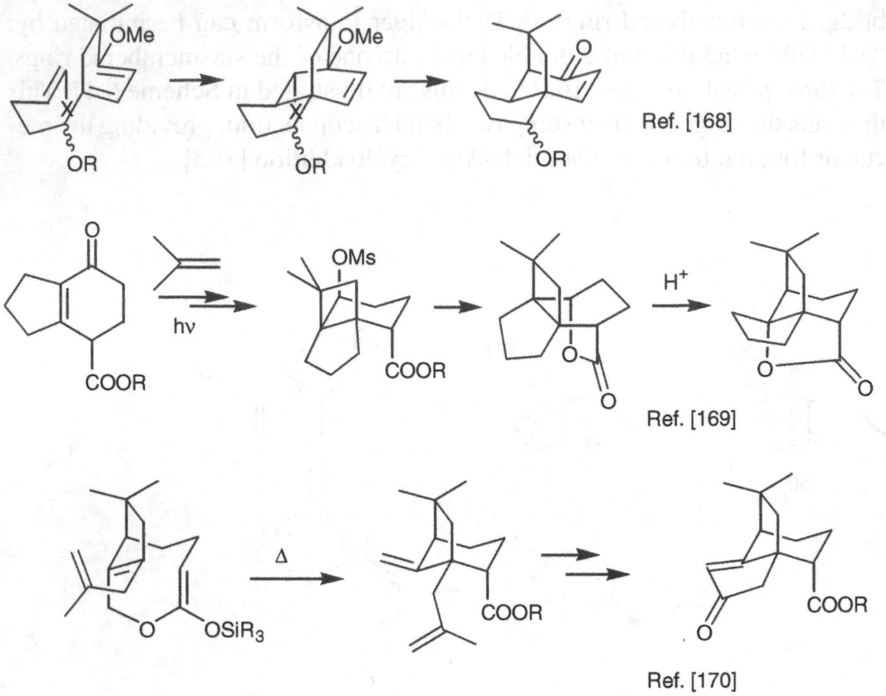

Ref. [168]

Ref. [169]

Ref. [170]

Scheme 6.48 Surprising routes to the quadrone skeleton enlisting backbone rearrangement reactions

6.3 "Overbred" Intermediates

In the synthesis of polycyclic compounds, one occasionally encounters an example in which intermediates are structurally more complex than the target to be reached. In these cases, the molecular skeleton is first "overbred," only to be later reduced in complexity. This seems illogical, unless one recognizes that bond cleavage in the forward synthetic direction may actually be a productive process. For instance, a bicyclization generates two skeletal bonds, but if one of them is not needed in the target structure, it may be cleaved in a subsequent step. Such removal of a "surplus" bond is exemplified in Scheme 6.49 [171].

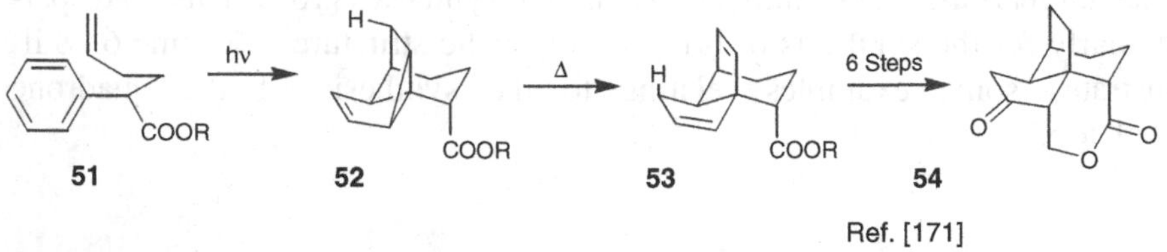

Ref. [171]

Scheme 6.49 Synthesis of the quadrone skeleton via an "overbred" intermediate

In this case, an intramolecular photocycloaddition between the phenyl ring and the pendant double bond in **51** led to tetracyclic intermediate **52**, which has one bond in excess of what is needed for quadrone. This excess bond is a cyclopropane bond, which may be severed in the course of a thermal 1,5-hydrogen shift, to give compound **53**. Model studies revealed that it is possible to introduce into **53** the decoration needed to obtain quadrone [171]. What should be emphasized here is that intermediate **52** is more complex than the quadrone skeleton in **54**. However, the ease of formation of intermediate **52** and the subsequent structural "correction" in only a single operation render this approach to quadrone via an overbred skeleton highly attractive. Scheme 6.50 provides a further example from studies aimed at quadrone [172, 173].

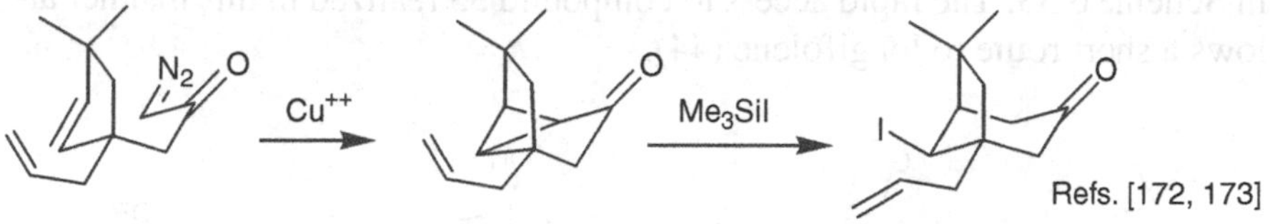

Refs. [172, 173]

Scheme 6.50 Synthesis of quadrone via an intermediate with an "overbred" skeleton

Planning reaction schemes which incorporate overbred molecular skeletons looks like an exotic undertaking. To do this, one needs a good knowledge of C-C bond breaking reactions [167, 174], which could be introduced retrosynthetically as an "add bond" operation. "Add bond" is used in the retrosynthesis of anellated and spirocyclic targets (Scheme 6.51).

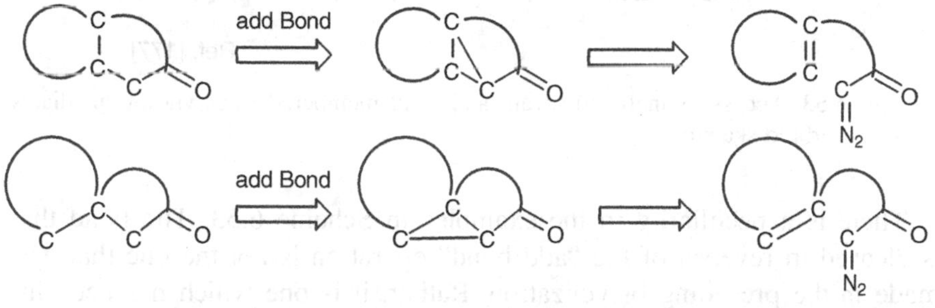

Scheme 6.51 Add bond strategy for syntheses of bicycles via an overbred skeleton

The forward execution of reaction sequences in which two bonds are made, and then one of them is cleaved in the next step, is well established (Scheme 6.52) [175].

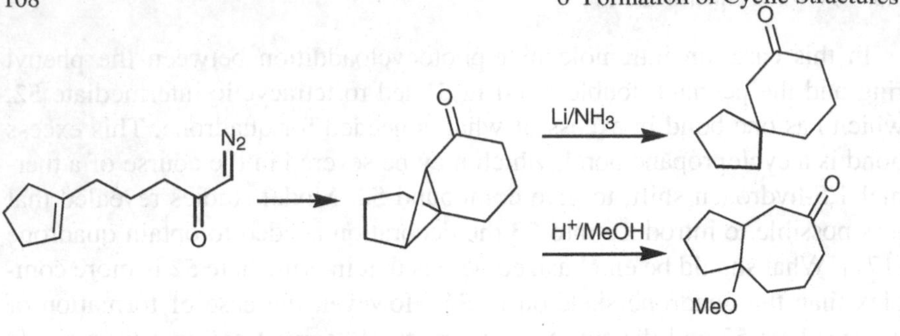

Scheme 6.52 Routes to anellated cycles and spirocycles via an "overbred" skeleton

Approaches via an overbred skeleton turn out to be very versatile for accessing seven- or eight-membered ring systems. Examples for both are given in Scheme 6.53. The rapid access to compound **55** realized in this manner allows a short route to longifolene (**44**).

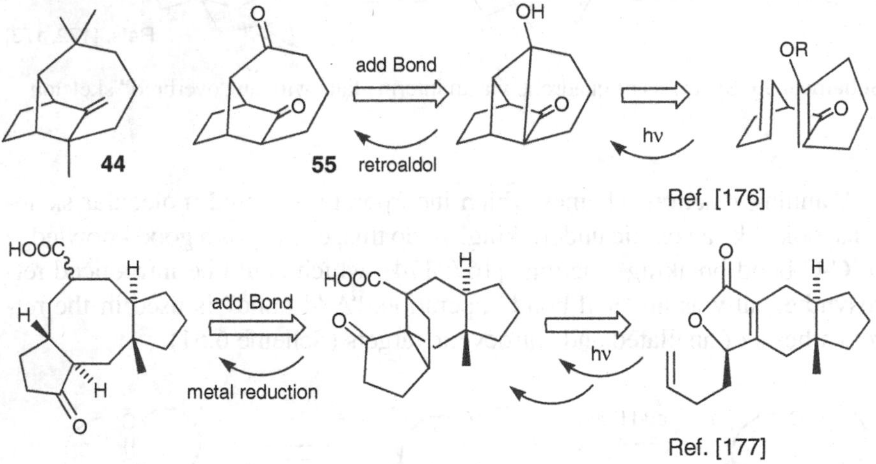

Scheme 6.53 Access to anellated seven- and eight-membered rings via intermediates with an overbred skeleton

There is a peculiarity in the examples in Scheme 6.53. The bond that is cleaved in reversal of the "add bond" operation is not the one that was made in the preceding bicyclization. Rather, it is one which has been intentionally introduced with one of the partners of the bicyclization. For the generation of eight-membered rings via an overbred skeleton, not only do bicyclo[4.2.0]octane systems serve well (Scheme 6.53), but so too do bicyclo[3.3.0]octane systems, i.e., anellated 5/5-ring systems (Scheme 6.54).

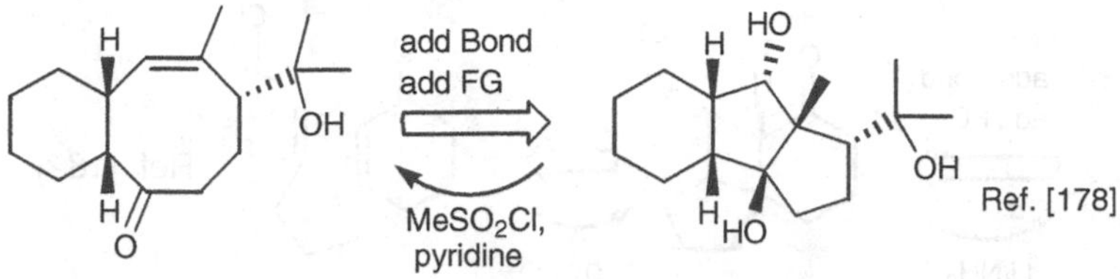

Scheme 6.54 Synthesis of an anellated eight-membered ring by cleavage of a 5/5 ring system

Targets possessing two adjacent cis positioned side chains on a ring lend themselves to strategies involving an overbred skeleton. Once you connect these side chains retrosynthetically by "add bond" (equivalent to "reconnect" in this case) to create another ring, you have an intermediate which could arise by cycloaddition to the original ring (Scheme 6.55). Even two neighboring methyl substituents on a ring might be generated using this protocol [179].

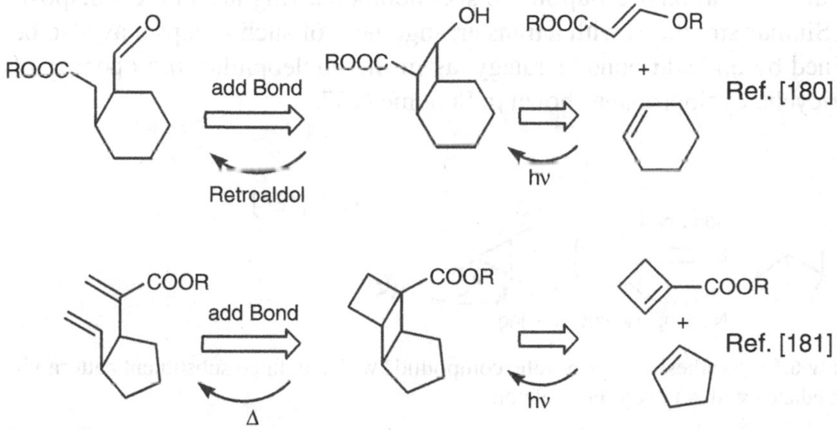

Scheme 6.55 Synthesis of cis-1,2-disubstituted rings via intermediates with an overbred skeleton

A reaction sequence involving an overbred skeleton becomes a prime choice when tackling a spirocyclic system with a stereochemically defined arrangement of substituents. Examples are given in Scheme 6.56.

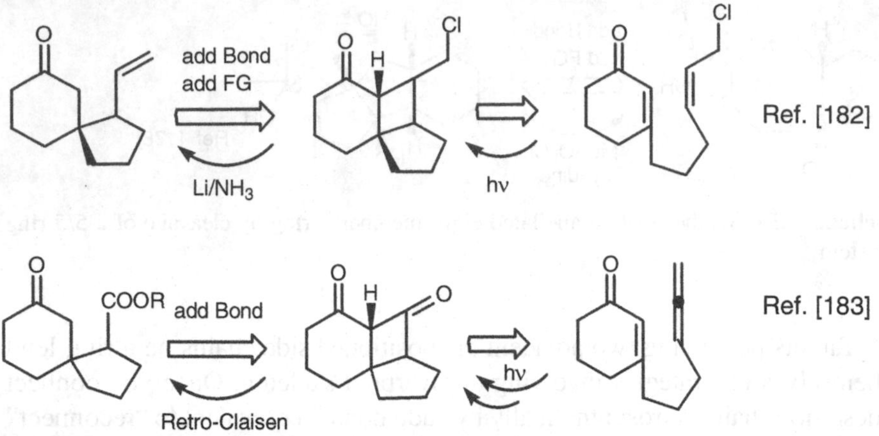

Scheme 6.56 Synthesis of spirocyclic compounds via intermediates with an overbred skeleton

In the examples in Scheme 6.56, the substituent on the five-membered ring and the carbonyl group in the six-membered ring are in a cis disposition. Similar structures with a trans arrangement of such groups may also be obtained by an "add bond" strategy, as in the nucleophilic ring opening of the tricyclic cyclopropane shown in Scheme 6.57.

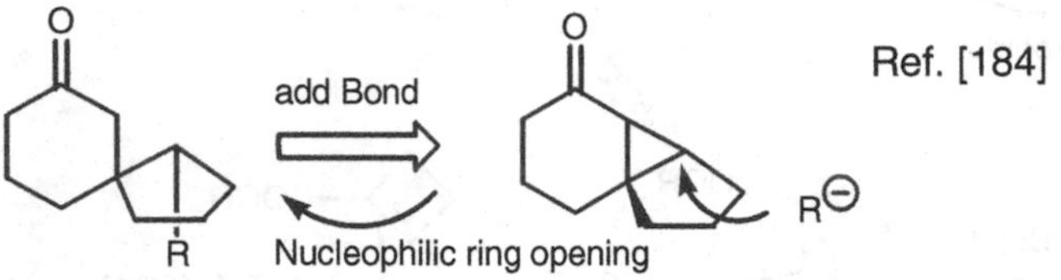

Scheme 6.57 Synthesis of spirocyclic compounds with a defined substituent pattern via intermediates with an overbred skeleton

Problems

6.1 1,5-Diaza-cis-decalin, the bicyclic analogue of tetramethyl-ethylenediamine, shown in Scheme 6.58, is an interesting (chiral) ligand [185, 186]. Is there a problem with the possible bicyclization approach shown?

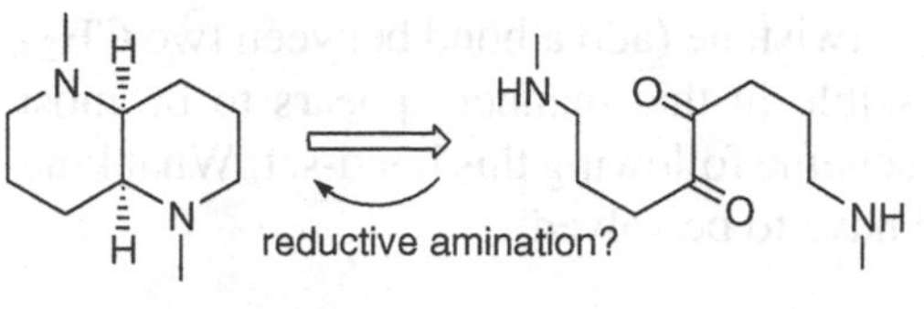

Scheme 6.58 Reductive amination in a bicyclization approach to 1,5-diaza-cis-decalin

For another synthesis of 1,5-diaza-cis-decalin, consider anellation approaches starting, e.g., from lysine.

In the end, a completely different solution [185, 186], which underscores one of the guidelines made for the synthesis of six-membered (carbo)cycles, turns out to be more viable.

6.2 Devise a retrosynthesis for the anellated tricycle shown in Scheme 6.59.

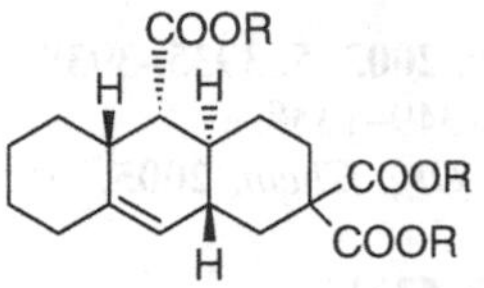

Scheme 6.59 Anellated tricycle

6.3 To conclude the discussion about planning syntheses of polycyclic target molecules, we may consider a rather simple symmetrical tricycle, tricyclo[3.3.1.1^{2,6}]decane, called twistane [187] (Scheme 6.60). As a consequence of its structural symmetry, twistane has only four different types of skeletal bonds. Can you identify them? Can you identify the most highly bridged ring?

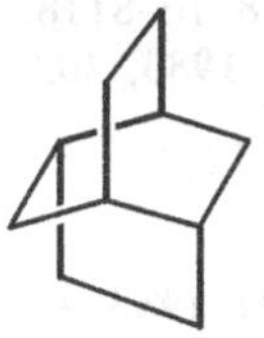

Scheme 6.60 Tricyclo[3.3.1.1^{2,6}]decane (twistane)

Which of the bonds in this ring marked as strategic would be optimal for a one-bond disconnection approach? Suggest a reaction scheme that would generate twistane according to that bond-set.

Consider a two-bond disconnection approach for twistane. What problem will you run into?

Apply the "add bond" strategy to twistane (add a bond between two CH_2-units). Which of the linkages possible in this manner appears to be most advantageous? Suggest a reaction scheme following this bond-set. What kind of regioselectivity problem would have to be solved?

References

1. C. P. Dell, *J. Chem. Soc., Perkin Trans. 1,* **1998**, 3873–3905.
2. P. A. Wender, J. A. Love, *Advances in Cycloaddition* **1999**, *5*, 1–45.
3. S. H. Bertz, *New J. Chem.* **2003**, *27*, 860–869.
4. G. Büchi, H. Wüest, *Helv. Chim. Acta* **1971**, *54*, 1767–1776.
5. M. Lautens, W. Klute, W. Tam, *Chem. Rev.* **1996**, *96*, 49–92.
6. S. J. Hedley, W. J. Moran, D. A. Price, J. P. A. Harrity, *J. Org. Chem.* **2003**, *68*, 4286–4292.
7. W. J. Moran, K. M. Goodenough, P. Raubo, J. P. A. Harrity, *Org. Lett.* **2003**, *5*, 3427–3429.
8. A. V. Kurdyumov, R. P. Hsung, K. Ihlen, J. Wang, *Org. Lett.* **2003**, *5*, 3935–3938.
9. J. P. A. Harrity, O. Provoost, *Org. Biomol. Chem.* **2005**, *3*, 1349–1358.
10. A. I. Gerasyuto, R. P. Hsung, N. Sydorenko, B. Slafer, *J. Org. Chem.* **2005**, *70*, 4248–4256.
11. R. Shintani, T. Hayashi, *J. Am. Chem. Soc.* **2006**, *128*, 6330–6331.
12. J. Sauer, R. Sustmann, *Angew. Chem., Int. Ed. Engl.* **1980**, *19*, 779–807. (*Angew. Chem.* **1980**, *92*, 773–801).
13. A. Hosomi, M. Saito, H. Sakurai, *Tetrahedron Lett.* **1980**, *21*, 355–358.
14. S. J. Danishefsky, *Acc. Chem. Res.* **1981**, *14*, 400–406.
15. J. Sauer, *Angew. Chem., Int. Ed. Engl.* **1967**, *6*, 16–33. (*Angew. Chem.* **1967**, *79*, 76–94).
16. P. G. Gassman, D. A. Singleton, J. J. Wilwerding, S. P. Chavan, *J. Am. Chem. Soc.* **1987**, *109*, 2182–2184.
17. P. G. Gassman, S. P. Chavan, *Tetrahedron Lett.* **1988**, *29*, 3407–3410.
18. M. Harmata, P. Rashatasakhon, *Tetrahedron* **2003**, *59*, 2371–2395.
19. B. M. Trost, J. Ippen, W. C. Vladuchick, *J. Am. Chem. Soc.* **1977**, *99*, 8116–8118.
20. B. M. Trost, W. C. Vladuchick, A. J. Bridges, *J. Am. Chem. Soc.* **1980**, *102*, 3554–3572.
21. P. Kozikowski, E. M. Huie, *J. Am. Chem. Soc.* **1982**, *104*, 2923–2925.
22. P. J. Proteau, P. B. Hopkins, *J. Org. Chem.* **1985**, *50*, 141–143.
23. P. V. Alston, M. D. Gordon, R. M. Ottenbrite, T. Cohen, *J. Org. Chem.* **1983**, *48*, 5051–5054.
24. D. A. Singleton, J. P. Martinez, *J. Am. Chem. Soc.* **1990**, *112*, 7423–7424.
25. D. A. Singleton, S.-W. Leung, *J. Org. Chem.* **1992**, *57*, 4796–4797.
26. S. Danishefsky, M. P. Prisbylla, S. Hiner, *J. Am. Chem. Soc.* **1978**, *100*, 2918–2920.
27. G. Stork, P. C. Tang, M. Casey, B. Goodman, M. Toyota, *J. Am. Chem. Soc.* **2005**, *127*, 16255–16262.
28. R. V. C. Carr, L. A. Paquette, *J. Am. Chem. Soc.* **1980**, *102*, 853–855.
29. O. DeLucchi, G. Modena, *Tetrahedron* **1984**, *40*, 2585–2632.
30. N. Ono, A. Kamimura, A. Kaji, *J. Org. Chem.* **1988**, *53*, 251–258.

31. A. C. Brown, L. A. Carpino, *J. Org. Chem.* **1985**, *50*, 1749–1750.
32. R. N. Warrener, R. A. Russel, R. Solomon, I. G. Pitt, D. N. Butler, *Tetrahedron Lett.* **1987**, *28*, 6503–6506.
33. G. Hilt, T. J. Korn, *Tetrahedron Lett.* **2001**, *42*, 2783–2785.
34. G. Hilt, K. I. Smolko, B. V. Lotsch, *Synlett* **2002**, 1081–1084.
35. S. Ranganathan, D. Ranganathan, A. K. Mehrotra, *Synthesis* **1977**, 289–296.
36. R. V. Williams, X. Lin, *J. Chem. Soc., Chem. Commun.* **1989**, 1872–1873.
37. V. K. Aggarwal, Z. Gültekin, R. S. Grainger, H. Adams, P. L. Spargo, *J. Chem. Soc., Perkin Trans. 1*, **1998**, 2771–2781.
38. W. G. Dauben, H. O. Krabbenhoft, *J. Org. Chem.* **1977**, *42*, 282–287.
39. R. W. Rickards, H. Rönneberg, *J. Org. Chem.* **1984**, *49*, 572–573.
40. G. M. Sammis, E. M. Flamme, H. Xie, D. M. Ho, E. J. Sorensen, *J. Am. Chem. Soc.* **2005**, *127*, 8612–8613.
41. M. J. Carter, I. Fleming, A. Percival, *J. Chem. Soc., Perkin Trans. 1*, **1981**, 2415–2434.
42. D. A. Evans, C. A. Bryan, C. L. Sims, *J. Am. Chem. Soc.* **1972**, *94*, 2891–2892.
43. E. Arce, M. C. Carreno, M. B. Cid, J. L. Garcia Ruano, *J. Org. Chem.* **1994**, *59*, 3421–3426.
44. E. J. Kantorowski, M. J. Kurth, *Tetrahedron* **2000**, *56*, 4317–4353.
45. Y. Mori, K. Yaegashi, T. Furukawa, *Tetrahedron* **1997**, *53*, 12917–12932.
46. J. W. Cornforth, R. Robinson, *J. Chem. Soc.* **1949**, 1855–1865.
47. M. Ramaiah, *Synthesis* **1984**, 529–570.
48. L. A. Paquette, *Topics Curr. Chem.* **1984**, *119*, 2–12.
49. M. Vandewalle, P. De Clercq, *Tetrahedron* **1985**, *41*, 1767–1831.
50. G. Büchi, M. Pawlack, *J. Org. Chem.* **1975**, *40*, 100–102.
51. K. G. Bilyard, P. J. Garratt, A. J. Underwood, R. Zahler, *Tetrahedron Lett.* **1979**, *20*, 1815–1818.
52. J. E. McMurry, D. D. Miller, *J. Am. Chem. Soc.* **1983**, *105*, 1660–1661.
53. M. E. Garst, B. J. McBride, A. T. Johnson, *J. Org. Chem.* **1983**, *48*, 8–16.
54. P. Wasnaire, T. de Merode, I. E. Markó, *Chem. Commun.* **2007**, 4755–4757.
55. R. J. Pariza, P. L. Fuchs, *J. Org. Chem.* **1983**, *48*, 2304–2306.
56. S. E. Denmark, J. P. Germanas, *Tetrahedron Lett.* **1984**, *25*, 1231–1234.
57. E. J. Corey, D. L. Boger, *Tetrahedron Lett.* **1978**, *19*, 2461–2464.
58. K. E. Harding, P. M. Puckett, J. L. Cooper, *Bioorg. Chem* **1978**, *7*, 221–234.
59. J. A. Thomas, C. H. Heathcock, *Tetrahedron Lett.* **1980**, *21*, 3235–3236.
60. W. P. Jackson, S. V. Ley, A. J. Whittle, *J. Chem. Soc., Chem. Commun.* **1980**, 1173–1174.
61. F. Näf, R. Decorzant, W. Thommen, *Helv. Chim. Acta* **1975**, *58*, 1808–1812.
62. J. M. Conia, G. Moinet, *Bull. Soc. Chim Fr.* **1969**, 500–508.
63. P. T. Lansbury, N. Nazarenko, *Tetrahedron Lett.* **1971**, *12*, 1833–1836.
64. Y. Hayakawa, K. Yokoyama, R. Noyori, *J. Am. Chem. Soc.* **1978**, *100*, 1799–1806.
65. B. M. Trost, D. P. Curran, *J. Am. Chem. Soc.* **1980**, *102*, 5699–5700.
66. H. J. Altenbach, *Angew. Chem., Int. Ed. Engl.* **1979**, *18*, 940–941. (*Angew. Chem.* **1979**, *91*, 1005–1006).
67. W. G. Dauben, D. J. Hart, *J. Org. Chem.* **1977**, *42*, 3787–3793.
68. G. Stork, A. Brizzolara, H. Landesman, J. Szmuszkovicz, R. Terrell, *J. Am. Chem. Soc.* **1963**, *85*, 207–222.
69. E. Piers, B. Abeysekera, J. R. Scheffer, *Tetrahedron Lett.* **1979**, *20*, 3279–3282.
70. M. Miyashita, T. Yanami, A. Yoshikoshi, *J. Am. Chem. Soc.* **1976**, *98*, 4679–4681.

71. P. T. Lansbury, E. J. Nienhouse, *J. Am. Chem. Soc.* **1966**, *88*, 4290–4291.
72. J. R. Matz, T. Cohen, *Tetrahedron Lett.* **1981**, *22*, 2459–2462.
73. M. Karpf, A. S. Dreiding, *Helv. Chim. Acta* **1979**, *62*, 852–865.
74. B. M. Trost, *Acc. Chem. Res.* **1974**, *7*, 85–92.
75. G. Stork, N. H. Baine, *J. Am. Chem. Soc.* **1982**, *104*, 2321–2323.
76. X. Shi, D. J. Gorin, F. D. Toste, *J. Am. Chem. Soc.* **2005**, *127*, 5802–5803.
77. A. Murfat, P. Helquist, *Tetrahedron Lett.* **1978**, *19*, 4217–4220.
78. W. C. Agosta, S. Wolff, *J. Org. Chem.* **1975**, *40*, 1699–1701.
79. E. Piers, C. K. Lau, I. Nagakura, *Tetrahedron Lett.* **1976**, *17*, 3233–3236.
80. S. Knapp, U. O'Connor, D. Mobilio, *Tetrahedron Lett.* **1980**, *21*, 4557–4560.
81. R. L. Danheiser, D. J. Carini, D. M. Fink, A. Basak, *Tetrahedron* **1983**, *39*, 935–947.
82. T. Hiyama, M. Shinoda, H. Nozaki, *Tetrahedron Lett.* **1978**, *19*, 771–774.
83. H. Stetter, I. Krüger-Hansen, M. Rizk, *Chem. Ber.* **1961**, *94*, 2702–2707.
84. E. Piers, V. Karunaratne, *J. Chem. Soc., Chem. Commun.* **1983**, 935–936.
85. N. N. Marinovic, H. Ramanathan, *Tetrahedron Lett.* **1983**, *24*, 1871–1874.
86. G. Majetich, R. Desmond, A. M. Casares, *Tetrahedron Lett.* **1983**, *24*, 1913–1916.
87. S. Danishefsky, S. Chackalamannil, B.-J. Uang, *J. Org. Chem.* **1982**, *47*, 2231–2232.
88. T. Shono, I. Nishiguchi, H. Omizu, *Chem. Lett.* **1976**, *5*, 1233–1236.
89. E. J. Corey, M. A. Tius, J. Das, *J. Am. Chem. Soc.* **1980**, *102*, 1742–1744.
90. J. Justicia, J. E. Oltra, J. M. Cuerva, *J. Org. Chem.* **2005**, *70*, 8265–8272.
91. J. Dijkink, W. N. Speckamp, *Tetrahedron Lett.* **1977**, *18*, 935–938.
92. D. Nasipuri, G. Das, *J. Chem. Soc., Perkin Trans. 1*, **1979**, 2776–2778.
93. T. Hudlicky, F. J. Koszyk, D. M. Dochwat, G. L. Cantrell, *J. Org. Chem.* **1981**, *46*, 2911–2915.
94. W. R. Roush, H. R. Gillis, A. I. Ko, *J. Am. Chem. Soc.* **1982**, *104*, 2269–2283.
95. E. Ciganek, *Org. Reactions* **1984**, *32*, 1–374.
96. D. Craig, *Chem. Soc. Rev.* **1987**, *16*, 187–238.
97. W. Oppolzer, *Angew. Chem., Int. Ed. Engl.* **1977**, *16*, 10–23. (*Angew. Chem.* **1977**, *89*, 10–24).
98. G. Brieger, J. N. Bennett, *Chem. Rev.* **1980**, *80*, 63–97.
99. M. E. Jung, K. M. Halweg, *Tetrahedron Lett.* **1981**, *22*, 2735–2738.
100. W. Oppolzer, W. Fröstl, H. P. Weber, *Helv. Chim. Acta* **1975**, *58*, 593–595.
101. M. Braun, *Nachr. Chem. Techn. Lab.* **1985**, *33*, 718–725.
102. N.-Y. Wang, C.-T. Hsu, C. J. Sih, *J. Am. Chem. Soc.* **1981**, *103*, 6538–6539.
103. N. N. Girotra, N. L. Wendler, *Tetrahedron Lett.* **1983**, *24*, 3687–3688.
104. P. A. Grieco, R. E. Zelle, R. Lis, J. Finn, *J. Am. Chem. Soc.* **1983**, *105*, 1403–1404.
105. S. J. Danishefsky, B. Simoneau, *J. Am. Chem. Soc.* **1989**, *111*, 2599–2604.
106. C. H. Heathcock, M. J. Taschner, T. Rosen, J. A. Tomas, C. R. Hadley, G. Popjak, *Tetrahedron Lett.* **1982**, *23*, 4747–4750.
107. P. M. Wovkulich, P. C. Tang, N. K. Chadha, A. D. Batcho, J. C. Barrish, M. R. Uskokovic, *J. Am. Chem. Soc.* **1989**, *111*, 2596–2599.
108. P. C. Anderson, D. L. J. Clive, C. F. Evans, *Tetrahedron Lett.* **1983**, *24*, 1373–1376.
109. R. L. Funk, C. J. Mossman, W. E. Zeller, *Tetrahedron Lett.* **1984**, *25*, 1655–1658.
110. M. Hirama, M. Uei, *J. Am. Chem. Soc.* **1982**, *104*, 4251–4253.
111. R. L. Funk, W. E. Zeller, *J. Org. Chem.* **1982**, *47*, 180–182.
112. E. A. Deutsch, B. B. Snider, *J. Org. Chem.* **1982**, *47*, 2682–2684.
113. G. E. Keck, D. F. Kachensky, *J. Org. Chem.* **1986**, *51*, 2487–2493.

114. Y.-L. Yang, S. Manna, J. R. Falck, *J. Am. Chem. Soc.* **1984**, *106*, 3811–3814.
115. J. R. Falck, Y.-L. Yang, *Tetrahedron Lett.* **1984**, *25*, 3563–3566.
116. T. Sammakia, D. M. Johns, G. Kim, M. A. Berliner, *J. Am. Chem. Soc.* **2005**, *127*, 6504–6505.
117. H. Muxfeldt, W. Rogalski, *J. Am. Chem. Soc.* **1965**, *87*, 933–934.
118. J. J. Korst, J. D. Johnston, K. Butler, E. J. Bianco, L. H. Conover, R. B. Woodward, *J. Am. Chem. Soc.* **1968**, *90*, 439–457.
119. K. Maruyama, H. Uno, Y. Naruta, *Chem. Lett.* **1983**, *12*, 1767–1770.
120. G. A. Kraus, S. H. Woo, *J. Org. Chem.* **1987**, *52*, 4841–4846.
121. Y. Naruta, Y. Nishigaichi, K. Maruyama, *J. Org. Chem.* **1988**, *53*, 1192–1199.
122. T. Kametani, H. Nemoto, H. Ishikawa, K. Shiroyama, H. Matsumoto, K Fukumoto, *J. Am. Chem. Soc.* **1977**, *99*, 3461–3466.
123. W. Oppolzer, K. Bättig, M. Petrzilka, *Helv. Chim. Acta* **1978**, *61*, 1945–1947.
124. W. Oppolzer, D. A. Roberts, *Helv. Chim. Acta* **1980**, *63*, 1703–1705.
125. K. C. Nicolaou, W. E. Barnette, P. Ma, *J. Org. Chem.* **1980**, *45*, 1463–1470.
126. G. Quinkert, U. Schwartz, H. Stark, W.-D. Weber, H. Baier, F. Adam, G. Dürner, *Angew. Chem., Int. Ed. Engl.* **1980**, *19*, 1029–1030. (*Angew. Chem.* **1980**, *92*, 1062–1063).
127. R. L. Funk, K. P. C. Vollhardt, *J. Am. Chem. Soc.* **1977**, *99*, 5483–5484.
128. E. J. Corey, W. J. Howe, H. W. Orf, D. A. Pensak, G. Petersson, *J. Am. Chem. Soc.* **1975**, *97*, 6116–6124.
129. B. Lei, A. G. Fallis, *J. Org. Chem.* **1993**, *58*, 2186–2195.
130. B. B. Snider, H. Lin, *J. Am. Chem. Soc.* **1999**, *121*, 7778–7786.
131. G. Scheffler, H. Seike, E. J. Sorensen, *Angew. Chem., Int. Ed.* **2000**, *39*, 4593–4596. (*Angew. Chem.* **2000**, *112*, 4783–4785).
132. M. Ousmer, N. A. Braun, M. A. Ciufolini, *Org. Lett.* **2001**, *3*, 765–767.
133. J.-H. Maeng, R. L. Funk, *Org. Lett.* **2001**, *3*, 1125–1128.
134. D. J. Wardrop, W. Zhang, *Org. Lett.* **2001**, *3*, 2353–2356.
135. K. M. Brummond, J. Lu, *Org. Lett.* **2001**, *3*, 1347–1349.
136. J. Sisko, J. R. Henry, S. M. Weinreb, *J. Org. Chem.* **1993**, *58*, 4945–4951.
137. S. Hong, J. Yang, S. M. Weinreb, *J. Org. Chem.* **2006**, *71*, 2078–2089.
138. C. S. Ge, S. Hourcade, A. Ferdenzi, A. Chiaroni, S. Mons, B. Delpech, C. Marazano, *Eur. J. Org. Chem.* **2006**, 4106–4114.
139. R. Downham, F. W. Ng, L. E. Overman, *J. Org. Chem.* **1998**, *63*, 8096–8097.
140. C. J. Douglas, S. Hiebert, L. E. Overman, *Org. Lett.* **2005**, *7*, 933–936.
141. D. J. Denhart, D. A. Griffith, C. H. Heathcock, *J. Org. Chem.* **1998**, *63*, 9616–9617.
142. M. E. Maier, *Nachr. Chem. Tech. Lab.* **1993**, *41*, 1120–1128.
143. D. A. Evans, C. H. Mitch, *Tetrahedron Lett.* **1982**, *23*, 285–288.
144. K. A. Parker, D. Fokas, *J. Am. Chem. Soc.* **1992**, *114*, 9688–9689.
145. K. A. Parker, D. Fokas, *J. Org. Chem.* **2006**, *71*, 449–455.
146. B. M. Trost, W. Tang, F. Dean Toste, *J. Am. Chem. Soc.* **2005**, *127*, 14785–14803.
147. C. Y. Hong, N. Kado, L. E. Overman, *J. Am. Chem. Soc.* **1993**, *115*, 11028–11029.
148. C. Y. Hong, L. E. Overman, *Tetrahedron Lett.* **1994**, *35*, 3453–3456.
149. D. Trauner, J. W. Bats, A. Werner, J. Mulzer, *J. Org. Chem.* **1998**, *63*, 5908–5918.
150. S. D. Burke, C. W. Murtiashaw, J. O. Saunders, J. A. Oplinger, M. S. Dike, *J. Am. Chem. Soc.* **1984**, *106*, 4558–4566.
151. S. Kim, D. H. Oh, J.-Y. Yoon, J. H. Cheong, *J. Am. Chem. Soc.* **1999**, *121*, 5330–5331.
152. A. B. Smith III, B. A. Wexler, J. Slade, *Tetrahedron Lett.* **1982**, *23*, 1631–1634.

153. A. S. Kende, B. Roth, P. J. Sanfilippo, T. J. Blacklock, *J. Am. Chem. Soc.* **1982**, *104*, 5808–5810.
154. S. Danishefsky, K. Vaughan, R. Gadwood, K. Tsuzuki, *J. Am. Chem. Soc.* **1981**, *103*, 4136–4141.
155. W. K. Bornack, S. S. Bhagwat, J. Ponton, P. Helquist, *J. Am. Chem. Soc.* **1981**, *103*, 4647–4648.
156. K. Kon, K. Ito, S. Isoe, *Tetrahedron Lett.* **1984**, *25*, 3739–3742.
157. A. P. Neary, P. J. Parsons, *J. Chem. Soc., Chem. Commun.* **1989**, 1090–1091.
158. R. H. Schlessinger, J. L. Wood, A. J. Poss, R. A. Nugent, W. H. Parsons, *J. Org. Chem.* **1983**, *48*, 1146–1147.
159. J. M. Dewanckele, F. Zutterman, M. Vandewalle, *Tetrahedron* **1983**, *39*, 3235–3244.
160. E. Piers, N. Moss, *Tetrahedron Lett.* **1985**, *26*, 2735–2738.
161. H.-J. Liu, M. Llinas-Brunet, *Can. J. Chem.* **1988**, *66*, 528–530.
162. C. H. Heathcock, *Angew. Chem., Int. Ed. Engl.* **1992**, *31*, 665–681. (*Angew. Chem.* **1992**, *104*, 675–691).
163. G. Büchi, W. D. MacLeod Jr., *J. Am. Chem. Soc.* **1962**, *84*, 3205–3206.
164. M. Dobler, J. D. Dunitz, B. Gubler, H. P. Weber, G. Büchi, O. J. Padilla, *Proc. Chem. Soc. (London)* **1963**, 383.
165. E. J. Corey, W. T. Wipke, *Science* **1969**, *166*, 178–192.
166. F. Näf, R. Decorzant, W. Giersch, G. Ohloff, *Helv. Chim. Acta* **1981**, *64*, 1387–1397.
167. J. B. Hendrickson, *J. Am. Chem. Soc.* **1986**, *108*, 6748–6756.
168. P. A. Wender, D. J. Wolanin, *J. Org. Chem.* **1985**, *50*, 4418–4420.
169. A. B. Smith III, J. P. Konopelski, B. A. Wexler, P. A. Spengeler, *J. Am. Chem. Soc.* **1991**, *113*, 3533–3542.
170. R. L. Funk, M. M. Abelman, *J. Org. Chem.* **1986**, *51*, 3247–3248.
171. P. A. Wender, *Selectivity – a Goal for Synthetic Efficiency* (Eds.: W. Bartmann, B. M. Trost), Verlag Chemie, Weinheim, **1984**, pp. 335–348.
172. T. Imanishi, M. Matsui, M. Yamashita, C. Iwata, *Tetrahedron Lett.* **1986**, *27*, 3161–3164.
173. T. Imanishi, M. Matsui, M. Yamashita, C. Iwata, *J. Chem. Soc., Chem. Commun.* **1987**, 1802–1804.
174. D. Caine, W. R. Pennington, T. L. Smith jr, *Tetrahedron Lett.* **1978**, 2663–2666.
175. J. F. Ruppert, J. D. White, *J. Am. Chem. Soc.* **1981**, *103*, 1808–1813.
176. W. Oppolzer, T. Godel, *J. Am. Chem. Soc.* **1978**, *100*, 2583–2584.
177. R. M. Coates, J. W. Muskopf, P. A. Senter, *J. Org. Chem.* **1985**, *50*, 3541–3557.
178. C. M. Amann, P. V. Fisher, M. L. Pugh, F. G. West, *J. Org. Chem.* **1998**, *63*, 2806–2807.
179. A. Hosomi, Y. Matsuyama, H. Sakurai, *J. Chem. Soc., Chem. Commun.* **1986**, 1073–1074.
180. B. A. Pearlman, *J. Am. Chem. Soc.* **1979**, *101*, 6398–6404.
181. P. A. Wender, J. C. Lechleiter, *J. Am. Chem. Soc.* **1977**, *99*, 267–268.
182. W. Oppolzer, *Acc. Chem. Res.* **1982**, *15*, 135–141.
183. D. Becker, Z. Harel, M. Nagler, A. Gillon, *J. Org. Chem.* **1982**, *47*, 3297–3306.

184. R. D. Clark, C. H. Heathcock, *Tetrahedron Lett.* **1975**, *16*, 529–532.
185. A. Gil de Oliveira Santos, W. Klute, J. Torode, V. P. W. Böhm, E. Cabrita, J. Runsink, R. W. Hoffmann, *New. J. Chem.* **1998**, 993–997.
186. X. Li, L. B. Schenkel, M. C. Kozlowski, *Org. Lett.* **2000**, *2*, 875–878.
187. D. P. G. Hannon, R. N. Young, *Austr. J. Chem.* **1976**, *29*, 145–161.

Chapter 7
Protecting Groups

Abstract Use of protecting groups documents our inability to do synthesis properly (in contrast to biosynthesis). The disadvantages that go along with the use of protecting groups can be minimized by a proper choice of short-term, medium-term, and long-term protecting groups, when in situ protection schemes and the use of latent functionality is not available.

During the synthesis of a complex target, synthetic organic chemists have grown accustomed to using protecting groups in order to shield particular functional groups from the threat of reagents and reaction conditions necessary for elaboration elsewhere in the molecule. In recent decades a whole arsenal of protecting groups for all conceivable functional groups has been developed [1, 2]. Scheme 7.1 illustrates the protecting group pattern of the

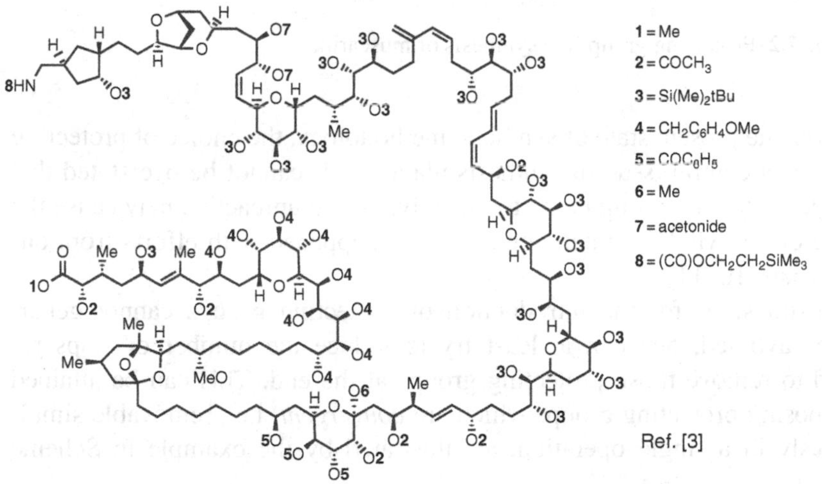

1 = Me
2 = COCH$_3$
3 = Si(Me)$_2$tBu
4 = CH$_2$C$_6$H$_4$OMe
5 = COC$_6$H$_5$
6 = Me
7 = acetonide
8 = (CO)OCH$_2$CH$_2$SiMe$_3$

Ref. [3]

Scheme 7.1 Protecting group pattern in the final phase of the synthesis of palytoxin carboxylic acid

R.W. Hoffmann, *Elements of Synthesis Planning*,
DOI 10.1007/978-3-540-79220-8_7, © Springer-Verlag Berlin Heidelberg 2009

epochal synthesis of palytoxin carboxylic acid [3], an enterprise that involved eight different types of protecting groups (42 in all). Each of these had to be introduced individually, and the whole flock had to be removed at the very end of the synthesis in five separate steps.

The universal (and frequently promiscuous) use of protecting groups is a telltale sign that chemists are not (yet) in a position to do synthesis right! Protecting groups normally require two operations (introduction and removal)—which, strictly speaking, are counterproductive, since they reduce the overall efficiency of the synthesis. While nature generates the most complex natural products without recourse to protecting groups, a protecting group free synthesis [4, 5, 6, 7] of even small multifunctional molecules, such as Fleet's synthesis of muscarine [8], cf. also [9] (Scheme 7.2), appears exceptional for most chemists.

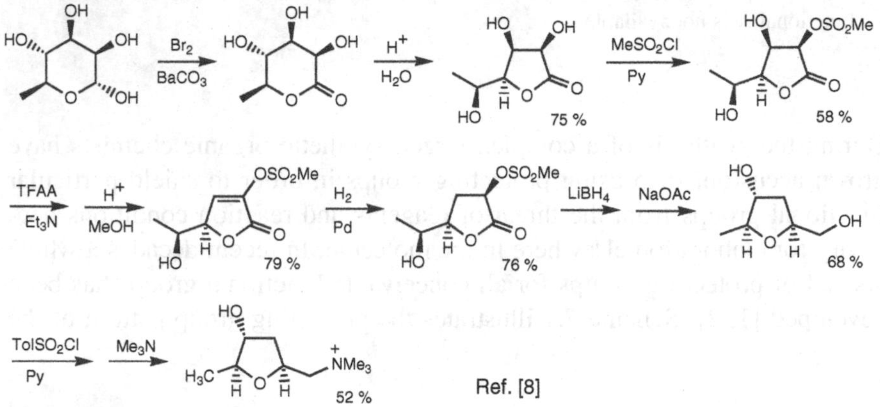

Scheme 7.2 Protecting group free synthesis of muscarine

Given the present state of synthetic methodology, the choice of protecting groups is a central issue in synthesis planning. It cannot be overstated that a single protecting group being too reactive or too unreactive may cause the failure of the whole synthetic endeavor, as happened with efforts from our own group [10, 11].

If extra steps for the introduction of protecting groups cannot generally be avoided, one can at least try to reduce the number of steps required to remove those protecting groups at the end. This can be attained by choosing protecting groups which are *convergent*, i.e., removable simultaneously in a single operation, as illustrated by the example in Scheme 7.3 [12].

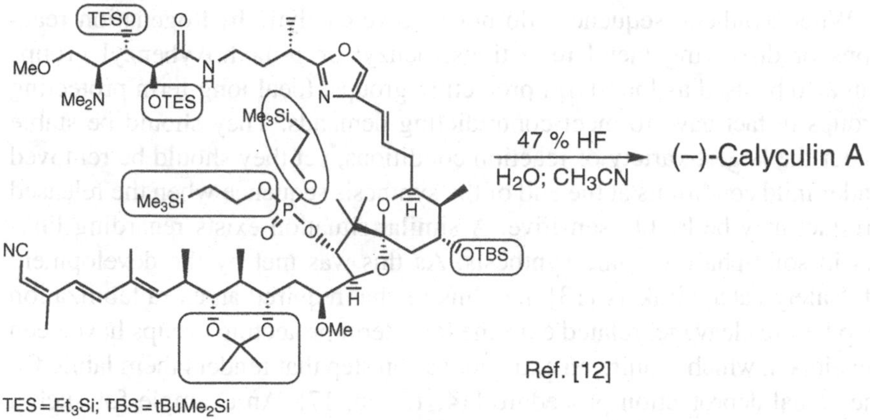

TES = Et₃Si; TBS = tBuMe₂Si

Scheme 7.3 Convergent pattern of protecting groups that can be removed in a single operation

Choosing protecting groups is a step that comes late in planning a synthesis, because it calls for detailed knowledge about the reaction steps and the conditions to be attempted in the synthetic sequence (including that of alternate routes!). Yet not all functional groups and attendant protecting groups have to endure all reaction steps. The earlier (or later) a protecting group is introduced in a synthetic sequence, the larger (or smaller) is the number of steps it has to go through unharmed. Accordingly, in planning protecting group patterns, one identifies:

- *long-term protecting groups*, which will be removed only at the end of the synthesis;
- *intermediate-term protecting groups*, which will be removed after a few steps;
- *short-term protecting groups*, which protect functionality for at most one or two steps.

The relationship between a short-term protecting group and a long-term protecting group is like that between a bandage and a cast. As long-term protecting groups have to endure the largest variety of reaction conditions, their removal requires special conditions, i.e., those that do not occur during normal synthetic operation. For this reason, silyl groups are frequently chosen as long-term protecting groups, removed at the end of the synthetic sequence by exposure to fluoride ions. See Scheme 7.4 for examples.

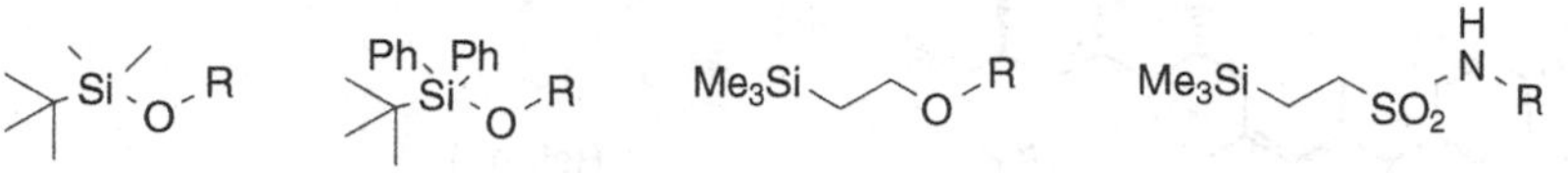

Scheme 7.4 Examples of long-term protecting groups

When synthetic sequences do not involve catalytic hydrogenation reactions or dissolving metal reductions, benzyl or *p*-methoxybenzyl groups can also be used as long-term protecting groups. Ideal long-term protecting groups in fact have to meet contradicting demands. They should be stable under the largest variety of reaction conditions, yet they should be removed under mild conditions at the end of the synthesis sequence when the released product may be highly sensitive. A similar situation exists regarding linkers in solid-phase organic synthesis. As this was met by the development of "safety-catch" linkers [13], i.e., linkers that required an extra labilization step before cleavage, related extreme long-term protecting groups have been developed, which require an extra activation step that renders them labile for the actual deprotection procedure [14, 15, 16, 17]. An example for such a carboxyl protecting group is given in Scheme 7.5.

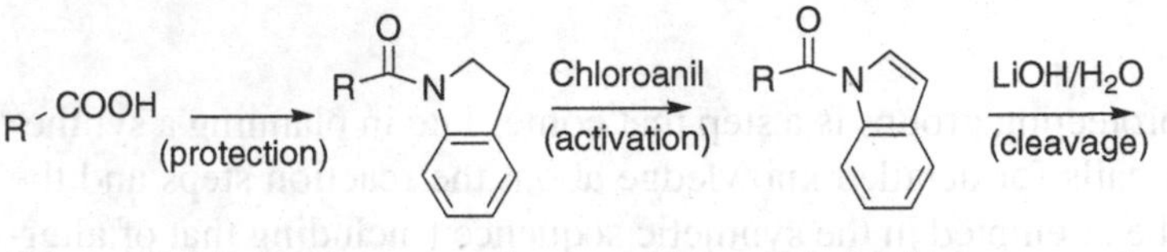

Scheme 7.5 Protecting group that has to be activated prior to removal

Intermediate-term and short-term protecting groups must be carefully chosen in order to guarantee the protection of the sensitive functionality for the intended number of steps, yet must also be removable under conditions that do not affect existing long-term protecting groups. Two protecting groups that may be introduced and removed independently of one another are said to be *orthogonal*. In Scheme 7.6, for example, the *p*-methoxybenzyl

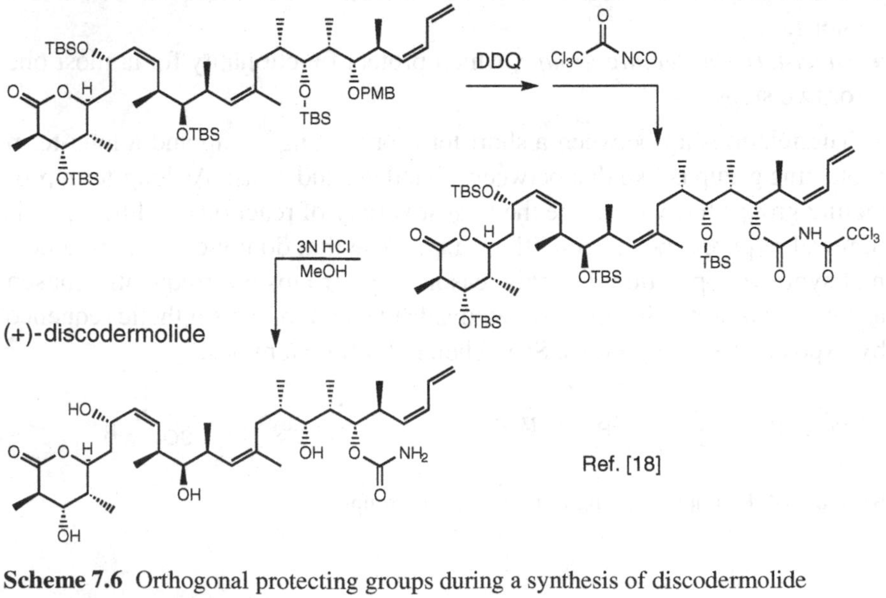

Scheme 7.6 Orthogonal protecting groups during a synthesis of discodermolide

group (PMB) is selectively cleaved in the presence of the long-term silyl protecting groups [18].

In this example, the carbamate functionality had been short-term protected as the trichloroacetate, which allows for convergent removal together with the long-term silyl protecting groups. One notes that the trichloroacetyl group had been brought in as part of the reagent trichloroacetyl isocyanate. The introduction of the protecting group, hence, did not require an extra step, nor did its later removal. In order to reduce the number of steps associated with protecting group management, it is useful to introduce the protecting group as part of a required reagent.

The number of available protecting groups differs for the various functional groups. There are many protecting groups for alcohols, but many less so for ketones [1]. This induces chemists to carry an ultimate ketone function through the synthetic sequence as a protected alcohol. At one stage or another the protected alcohol has to be deprotected and oxidized to a ketone. Rather than having two separate steps to achieve this, it is advantageous to use a protecting group that on deprotection effects simultaneous oxidation of the alcohol [19, 20]. An application is illustrated in Scheme 7.7 [21].

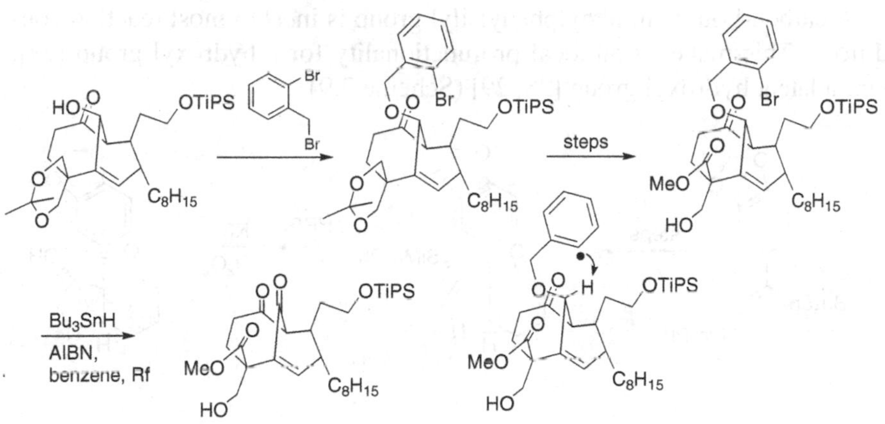

Scheme 7.7 Deprotection and simultaneous oxidation of an alcohol via a free radical 1,5-hydrogen shift.

In this case a free radical chain reaction is used, in which the tin radical abstracts the bromine atom. The resulting phenyl radical induces a 1,5-hydrogen shift. The resulting radical fragments, producing the ketone and a benzyl radical that carries the chain. The reaction sequence is remarkable because no free alcohol is generated as an intermediate, which in certain instances would be incompatible with other functionalities in the molecule. Thus, the ortho-bromo-benzyl ether serves as a direct latent ketone function.

Rather than introducing a functional group first and protecting it in a subsequent step, it is better to introduce the functional group initially in a latent (already protected) form [22, 23]. The example in Scheme 7.8 illustrates how a furan ring serves as a latent ester group to be unveiled later in the synthetic sequence [24]. Likewise an oxazole ring serves in an excellent manner as a latent carboxyl function [25, 26].

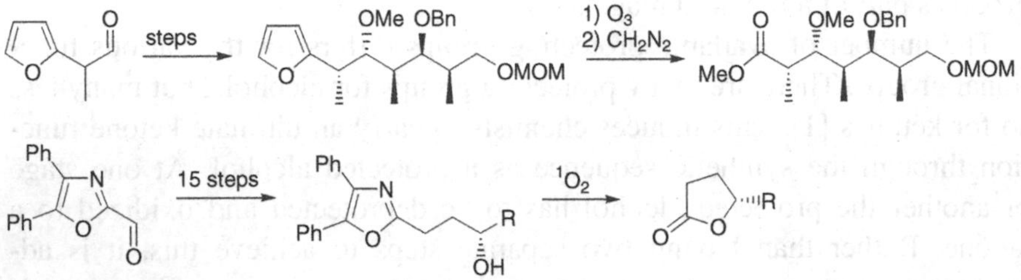

Scheme 7.8 A furan or an oxazole residue are synthetic equivalents of an ester function; they serve as a latent ester group

A carbon-bound dimethylphenylsilyl group is inert to most reaction conditions. This makes it an ideal profunctionality for a hydroxyl group [27], i.e., a latent hydroxyl group [28, 29] (Scheme 7.9).

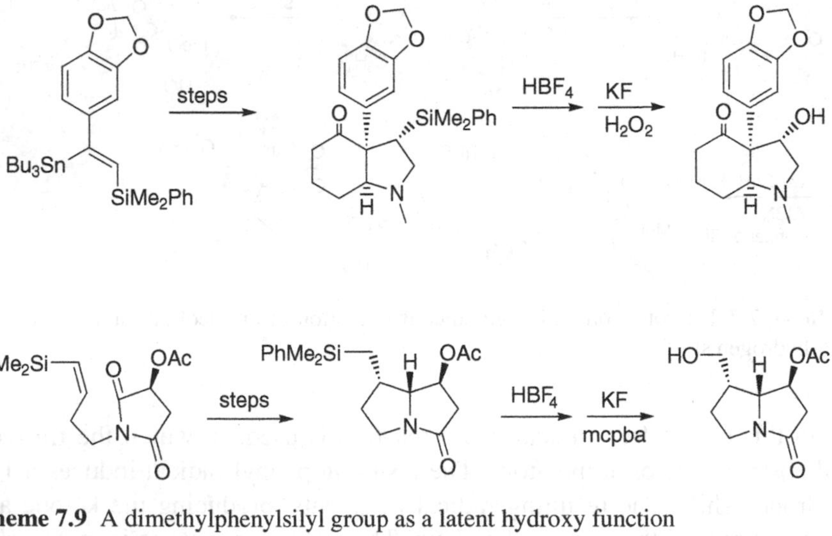

Scheme 7.9 A dimethylphenylsilyl group as a latent hydroxy function

Methoxyphenyl groups [30] and 2-alkylpyridines [31] may be carried un-harmed through long sequences of steps. At the end, their capacity as a latent cyclohexenone unit may be unveiled (Scheme 7.10).

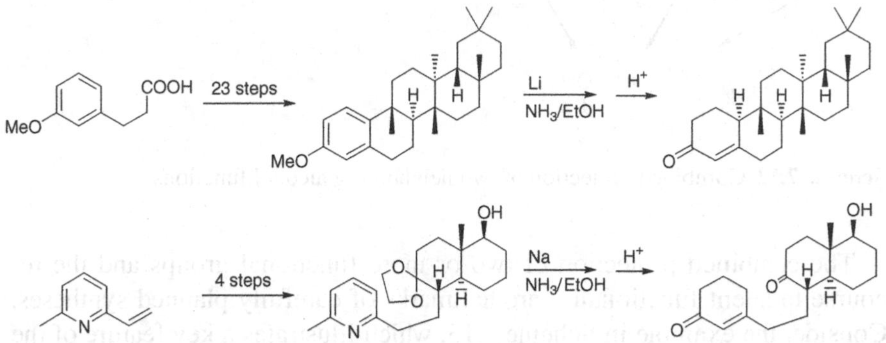

Scheme 7.10 Methoxyphenyl group, respectively, α-picolyl group as a latent cyclohexenone unit

A methoxyphenyl residue has also been utilized as a latent β-ketoester in the course of a synthetic sequence [32] (Scheme 7.11).

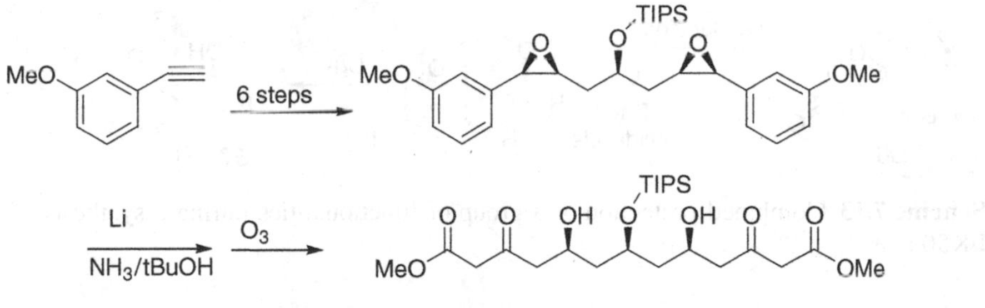

Scheme 7.11 Methoxyphenyl group as latent β-ketoester

Use of the methoxyphenyl group as a latent β-ketoester represents the combined protection of two functional groups (ketone and ester) in a single moiety. It is very common to combine the protection of neighboring alcohol functions as benzylidene acetal, as an acetonide, or as a siladioxane (Scheme 7.12). Of these, the benzylidene acetals are most versatile, as they allow the selective deprotection of either the sterically more encumbered or less encumbered hydroxyl function [1, 2].

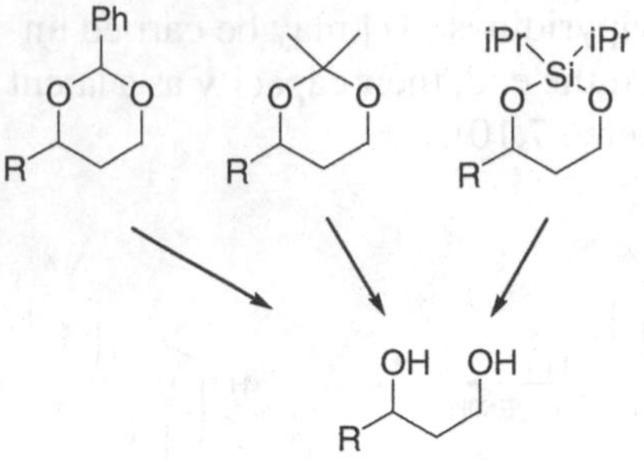

Scheme 7.12 Combined protection of two neighboring alcohol functions

The combined protection of two or more functional groups and the recourse to latent functionality are hallmarks of carefully planned syntheses. Consider the example in Scheme 7.13, which illustrates a key feature of the synthesis of FK506 by the Ireland group [33]. Only a partial structure of FK506 is shown. This unit, **56**, is the latent counterpart of the hydroxyl-, keto-, and alkene-functions present in **57**. The entity **56** was created early in the synthesis and carried through several steps until the desired functionality **57** was unveiled near the end of the synthesis.

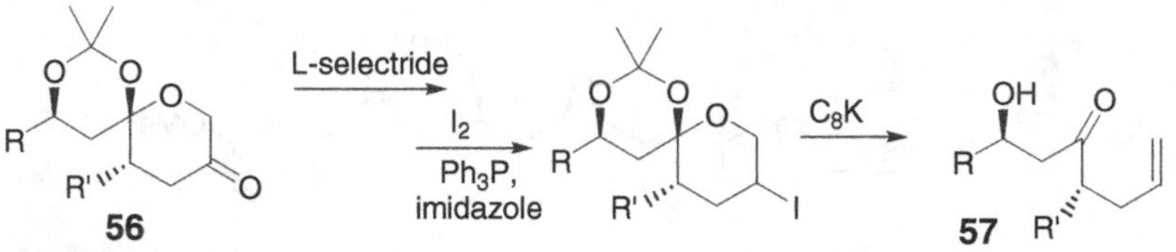

Scheme 7.13 Combined protection of a group of functionalities during a synthesis of FK506

The use of short-term protecting groups during a synthesis is as annoying as paying a 24-hour parking fee when one just wants to drink a cup of coffee. The effort associated with the introduction and removal of a short-term protecting group may be reduced when one manages to introduce and remove the protecting group in situ, i.e., without the isolation of any intermediates. The technique of in situ protection has been developed best for the protection of aldehyde groups in the presence of keto functionality. To this end, a metal amide is added to the (more reactive) aldehyde function. The resulting adduct is not susceptible to nucleophilic attack. One carries out the desired transformation on the keto group, and upon aqueous workup the aldehyde group is liberated (Scheme 7.14).

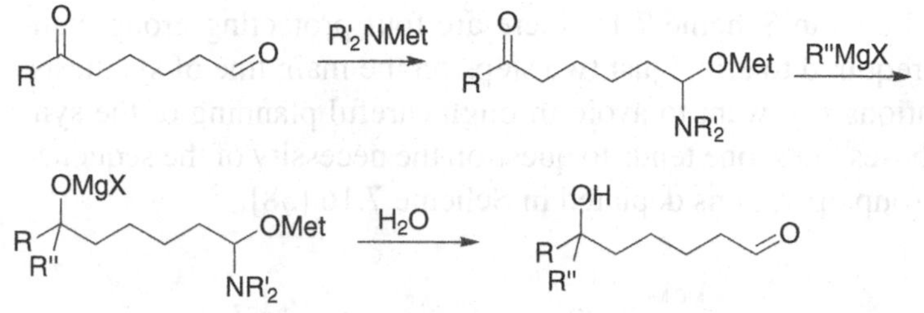

Met = Li, Ref. [34]; Me$_2$Al, Ref. [35]; (R'$_2$N)$_3$Ti, Ref. [36]

Scheme 7.14 In situ protection of an aldehyde function by addition of a metal amide

It remains to develop effective in situ protection tactics for other functional groups as well.

The principle of in situ protection is that the more reactive of two similar functional groups is (temporarily) transformed into a less reactive or unreactive moiety. However, the situation is much more complex if one wants to protect the less reactive of two functionalities. Usually one resorts to a (highly unsatisfactory) multistep protecting group dance [37] (Scheme 7.15).

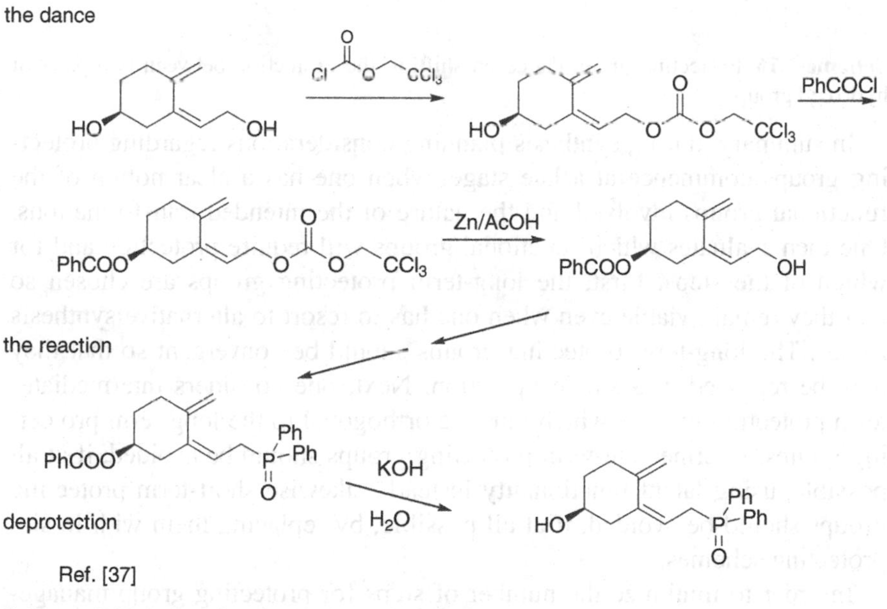

the dance

the reaction

deprotection

Ref. [37]

Scheme 7.15 Protecting group dance in order to protect the less reactive of two similar functional groups

In the case shown in Scheme 7.15, there are four protecting group management steps required to effect just two steps on the main line of synthesis. These are situations you want to avoid through careful planning of the synthesis. Along theses lines, one tends to question the necessity of the sequence of protecting group operations depicted in Scheme 7.16 [38].

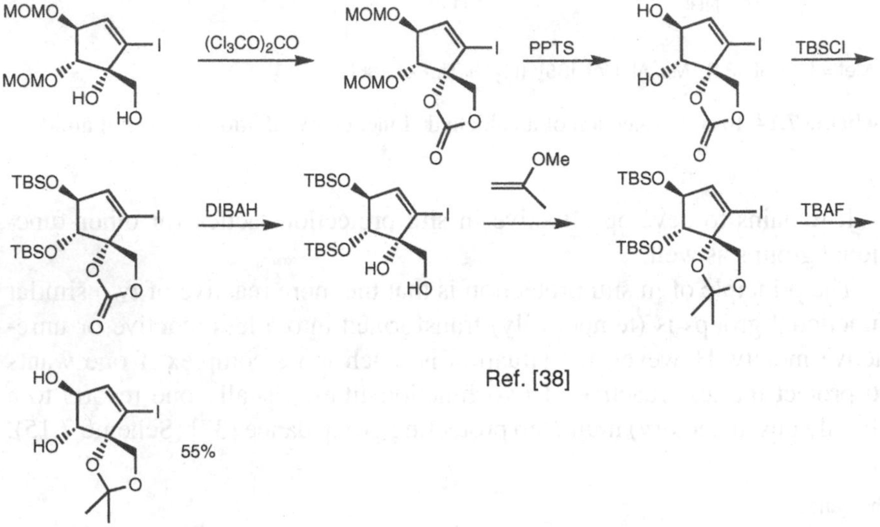

Ref. [38]

Scheme 7.16 Protecting group dance on shifting the protection between two pairs of hydroxyl groups

In summary, during synthesis planning considerations regarding protecting groups commence at a late stage, when one has a clear notion of the functional groups involved and the nature of the intended transformations. One then evaluates which functional groups will require protection and for which of the steps. First, the long-term protecting groups are chosen so that they remain viable even when one has to resort to alternative synthesis routes. The long-term protecting groups should be convergent so that they may be removed in a single operation. Next, one considers intermediate-term protecting groups, which must be orthogonal to the long-term protecting groups. Intermediate-term protecting groups should be avoided, if at all possible, using latent functionality instead. Likewise short-term protecting groups should be avoided, if at all possible, by replacing them with in situ protecting schemes.

In order to minimize the number of steps for protecting group management, check whether a different order of carrying out the construction steps could render some of the protecting groups unneccessary [39]. Check the possibility of combining several protecting groups into one! Make sure that the protecting group strategy does not contain any compromises or question marks. Nothing is more frustrating than to see a synthesis effort fail just because the protecting group pattern was inadequate [11, 40].

Problems

7.1 Suggest a protective group scheme (PG1–PG3) which is compatible with the following transformations [41]:

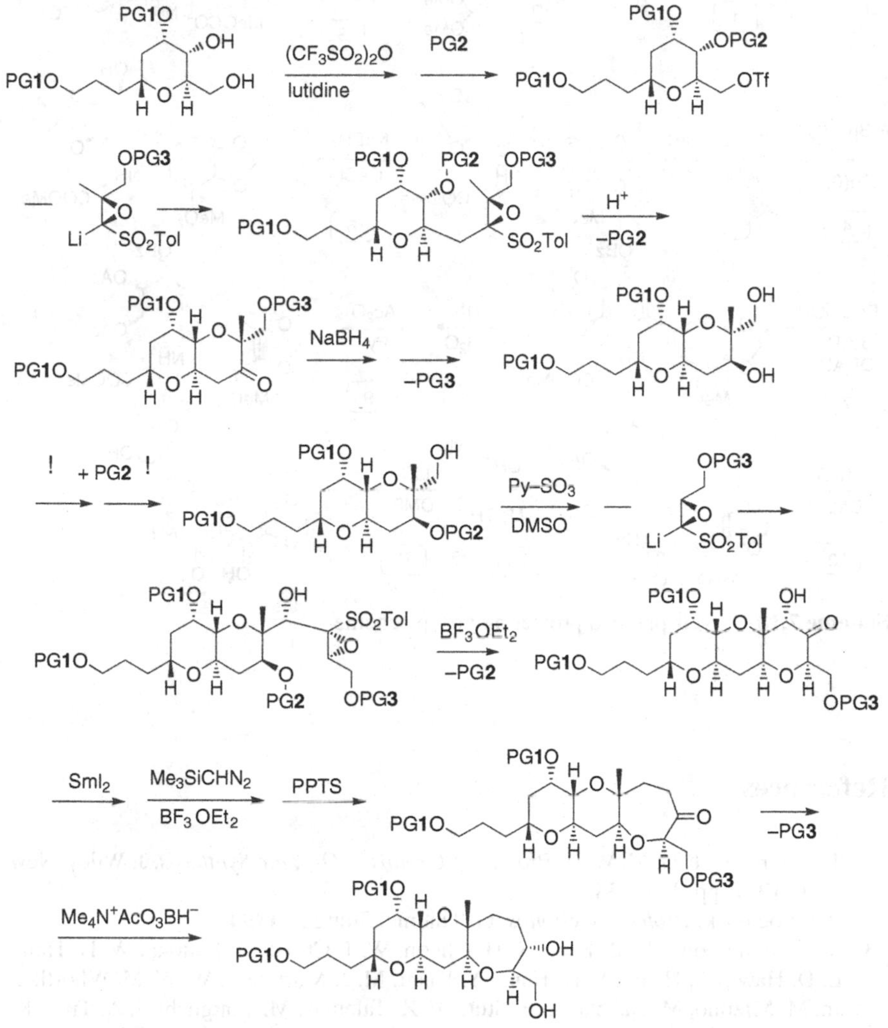

Scheme 7.17 Elaboration of a protecting group scheme

7.2 Try to rationalize the following protecting group regime of a narciclasine synthesis [42]:

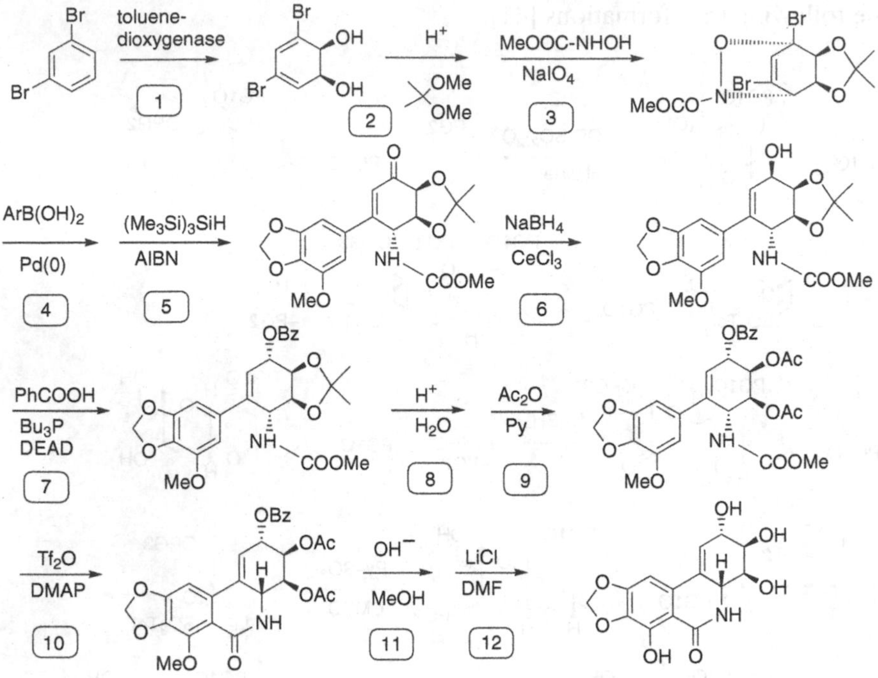

Scheme 7.18 A well-planned protecting group scheme

References

1. T. W. Greene, P. G. M. Wuts, *Protective Groups in Organic Synthesis*, J. Wiley, New York, **1991**, pp. 127–134.
2. P. J. Kocienski *Protecting Groups*, G. Thïeme, Stuttgart **1994**.
3. R. W. Armstrong, J.-M. Beau, S. H. Cheon, W. J. Christ, H. Fujioka, W.-H. Ham, L. D. Hawkins, H. Jin, S. H. Kang, Y. Kishi, M. J. Martinelli, W. W. McWhorther, Jr., M. Mizuno, M. Nakata, A. E. Stutz, F. X. Talamas, M. Taniguchi, J. A. Tino, K. Ueda, J. Uenishi, J. B. White, M. Yonaga, *J. Am. Chem. Soc.* **1989**, *111*, 7530–7533.
4. Y. Yokoyama, H. Hikawa, M. Mitsuhashi, A. Uyama, Y. Hiroki, Y. Murakami, *Eur. J. Org. Chem.* **2004**, 1244–1253.
5. P. S. Baran, J. M. Richter, *J. Am. Chem. Soc.* **2004**, *126*, 7450–7451.
6. P. S. Baran, J. M. Richter, *J. Am. Chem. Soc.* **2005**, *127*, 15394–15396.
7. Y. Zeng, J. Aubé, *J. Am. Chem. Soc.* **2005**, *127*, 15712–15713.
8. S. J. Mantell, G. W. J. Fleet, D. Brown, *J. Chem. Soc., Perkin Trans. 1*, **1992**, 3023–3027.

9. S. C. Archibald, R. W. Hoffmann, *Chemtracts-Org.Chem.* **1993**, *6*, 194–197.
10. R. Stürmer, R. W. Hoffmann, *Chem. Ber.* **1994**, *127*, 2519–2526.
11. G. Dahmann, R. W. Hoffmann, *Liebigs Ann. Chem.* **1994**, 837–845.
12. N. Tanimoto, S. W. Gerritz, A. Sawabe, T. Noda, S. A. Filla, S. Masamune, *Angew. Chem. Int. Ed. Engl.* **1994**, *33*, 673–675. (*Angew. Chem.* **1994**, *106*, 674–677).
13. P. Heidler, A. Link, *Bioorg. Med. Chem.* **2005**, *13*, 585–599.
14. K. C. Nicolaou, P. S. Baran, Y.-L. Zhong, K. C. Fong, Y. He, W. H. Yoon, H.-S. Choi, *Angew. Chem. Int. Ed. Engl.* **1999**, *38*, 1669–1675. (*Angew. Chem.* **1999**, *111*, 1781–1784).
15. O. J. Plante, S. L. Buchwald, P. H. Seeberger, *J. Am. Chem. Soc.* **2000**, *122*, 7148–7149.
16. H. Waldmann, H. Kunz, *J. Org. Chem.* **1988**, *53*, 4172–4175.
17. M. T. Crimmins, C. A. Carroll, A. J. Wells, *Tetrahedron Lett.* **1998**, *39*, 7005–7008.
18. A. B. Smith III, M. D. Kaufman, T. J. Beauchamp, M. J. LaMarche, H. Arimoto, *Org. Lett.* **1999**, *1*, 1823–1826.
19. V. Rukachasisirikul, U. Koert, R. W. Hoffmann, *Tetrahedron* **1992**, *48*, 4533–4544
20. D. P. Curran, H. Yu, *Synthesis* **1992**, 123–127
21. K. C. Nicolaou, Y. He, K. C. Fong, W. H. Yoon, H.-S. Choi, Y.-L. Zhong, P. S. Baran, *Org. Lett.* **1999**, *1*, 63–66.
22. D. Lednicer, *Adv. Org. Chem.* **1972**, *8*, 179–293.
23. L. Call, *Chem. i. u. Zeit* **1978**, *12*, 123–133.
24. G. Schmid, T. Fukuyama, K. Akasaka, Y. Kishi, *J. Am. Chem. Soc.* **1979**, *101*, 259–260.
25. H. H. Wasserman, R. J. Gambale, M. J. Pulwer, *Tetrahedron Lett.* **1981**, *22*, 1737–1740.
26. D. A. Evans, P. Nagorny, D. J. Reynolds, K. J. McRae, *Angew. Chem., Int. Ed.* **2007**, *46*, 541–544. (*Angew. Chem.* **2007**, *119*, 547–550).
27. G. R. Jones, Y. Landais, *Tetrahedron* **1996**, *52*, 7599–7662.
28. L. E. Overman, H. Wild, *Tetrahedron Lett.* **1989**, *30*, 647–650.
29. J. A. Dener, D. J. Hart, *Tetrahedron* **1988**, *44*, 7037–7041.
30. R. E. Ireland, M. I. Dawson, S. C. Welch, A. Hagenbach, J. Bordner, B. Trus, *J. Am. Chem. Soc.* **1973**, *95*, 7829–7841.
31. S. Danishefsky, P. Cain, A. Nagel, *J. Am. Chem. Soc.* **1975**, *97*, 380–387.
32. Z. Wang, D. Deschenes, *J. Am. Chem. Soc.* **1992**, *114*, 1090–1091.
33. R. E. Ireland, J. L. Gleason, L. D. Gegnas, T. K. Highsmith, *J. Org. Chem.* **1996**, *61*, 6856–6872.
34. D. L. Comins, *Synlett* **1992**, 615–625.
35. K. Maruoka, Y. Araki, H. Yamamoto, *Tetrahedron Lett.* **1988**, *29*, 3101–3104.
36. M. T. Reetz, B. Wenderoth, R. Peter, *J. Chem. Soc., Chem. Commun.* **1983**, 406–408.
37. B. Lythgoe, M. E. N. Nambudiry, J. Tideswell, *Tetrahedron Lett.* **1977**, *18*, 3685–3688.
38. Y. Koyama, M. J. Lear, F. Yoshimura, I. Ohashi, T. Mashimo, M. Hirama, *Org. Lett.* **2005**, *7*, 267–270.
39. K. J. Fraunhoffer, D. A. Bachovchin, M. C. White, *Org. Lett.* **2005**, *7*, 223–226.
40. T. Lister, M. V. Perkins, *Org. Lett.* **2006**, *8*, 1827–1830.
41. Y. Mori, K. Yaegashi, H. Furukawa, *J. Org. Chem.* **1998**, *63*, 6200–6209.
42. T. Hudlicky, U. Rinner, D. Gonzalez, H. Akgun, S. Schilling, P. Siengalewicz, T. A. Martinot, G. R. Pettit, *J. Org. Chem.* **2002**, *67*, 8726–8743.

Chapter 8
Ranking of Synthesis Plans

Abstract Hendrickson's definition of the "ideal synthesis" serves as a benchmark to assess synthesis plans. Criteria such as convergency, increase in complexity, and robustness are presented to rank synthesis plans and to pinpoint weaknesses therein.

Any ranking of plans for the synthesis of a given target compound depends on benchmarks which must be defined. Possible criteria may be

- the shortest route (time involved),
- the cheapest route (cost of materials),
- the novelty of the route (patentability),
- the greenest route (avoidance of problematic waste),
- the healthiest route (avoidance of toxic intermediates and side products),
- the most reliable route (lowest risk approach).

Aside from these external criteria, ranking of synthesis proposals could also follow systematic criteria, e.g., the step count. A synthesis that reaches the target in fewer steps than another one is considered superior. Every synthesis consists of obligatory steps, i.e., those by which the skeleton is made. When focusing on this aspect, the bond-set would give a lower limit to the number of steps involved in a projected synthesis, because any refunctionalization steps and protecting group management steps count in addition to the skeleton forming steps. Because of this, the bond-set does not reveal too much about a step count and the quality of a projected synthesis. For instance, the differences between the bond-sets of Woodward's [1] and Muxfeldt's [2] tetracycline syntheses are minimal (Scheme 8.1). Comparison of the bond-sets does not reveal that in Muxfeldt's synthesis three bonds are formed in one operation rendering this synthesis significantly shorter—22 steps in Woodward's synthesis versus 17 in Muxfeldt's.

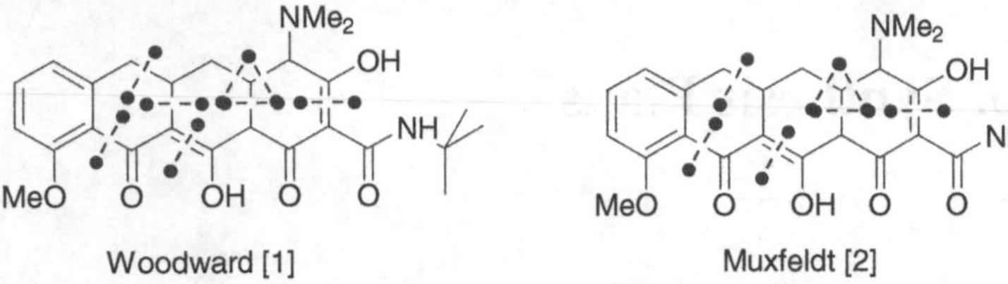

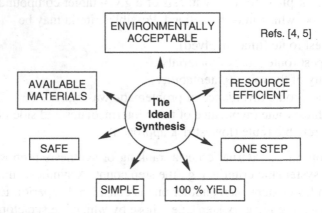

Scheme 8.1 Bond-sets of Woodward's and Muxfeld's syntheses of tetracycline

A comparison of the number of bonds in the bond-set with the total step count reveals that refunctionalization operations constitute the lion's share of the steps. When the step count is considered as a decisive criterion to judge the quality of a synthesis, then it becomes time to discuss some comments regarding the "ideal synthesis" (Scheme 8.2).

An "ideal synthesis" has been postulated by Turner [3] and by Wender [4, 5] as a one-pot reaction in which all starting materials are mixed, leading directly to the final product.

Scheme 8.2 Wender's comments regarding an ideal synthesis

"An ideal synthesis is generally regarded as one in which the target molecule is prepared from readily available, inexpensive starting materials in one simple, safe, environmentally acceptable, and resource-efficient operation that proceeds quickly and in quantitative yield."

Given the present imperfection of one-pot syntheses [6], this definition of the "ideal synthesis" is utopian, at least for in vitro chemical synthesis. One has to admit, though, that the in vivo biosynthesis of natural products in a living cell by and large fulfills the criterion of a so-defined ideal synthesis.

I would prefer a more realistic definition of the ideal synthesis, one from which guidelines for the planning of syntheses can be derived. This is the definition by Hendrickson [7]:

"Ideal Synthesis: The ideal synthesis creates a complex molecule ... in a sequence of only construction reactions involving no intermediary refunctionalizations, leading directly to the target, not only its skeleton but also its correctly placed functionality."

The reasoning of Hendrickson [7] is that only the skeletal bond-forming steps are obligatory in a synthesis. The superfluous refunctionalization steps could be dispensed with, once one succeeds at generating in the skeletal bond-forming reactions exactly that functionality which is present in the target or which is required to carry out the next skeletal bond-forming reaction. According to this reasoning, a numerical benchmark for the quality of a synthesis could be the ratio of the skeletal bond-forming reaction steps to that of the refunctionalization steps. For the tetracycline syntheses shown in Scheme 8.1, the ratios come out to be $6/16 = 0.37$ (Woodward) and $5/12 = 0.41$ (Muxfeldt). There are indeed syntheses reaching much higher numerical values for this ratio, e.g., 2.5 for Rawal's synthesis of geissoschicine [8] (Scheme 8.3).

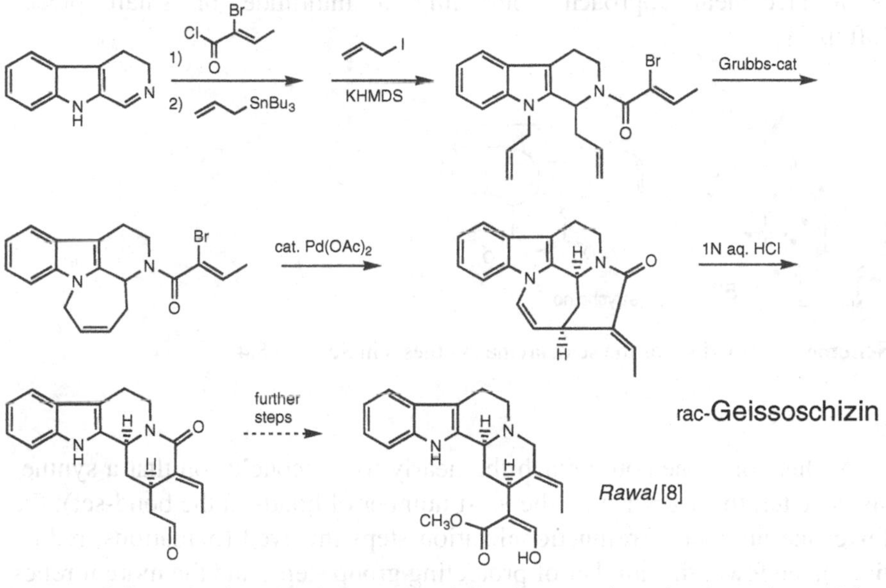

Scheme 8.3 Predominately skeleton bond-forming reactions in the geissoschicine synthesis by Rawal

A synthesis does not necessarily become good simply because it contains many skeletal bond-forming steps. This can be seen from an inspection of the setoclavine synthesis (ratio 1.67) [9] shown in Scheme 8.4.

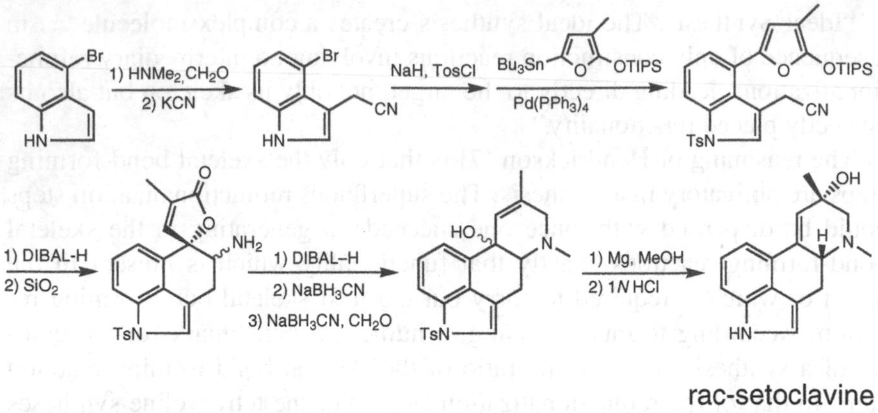

Scheme 8.4 Predominately skeletal bond-forming reactions in setoclavine synthesis

The bond set of this setoclavine synthesis (**58**, Scheme 8.5) reveals that it is a piecemeal approach combining a multitude of small pieces (often C_1).

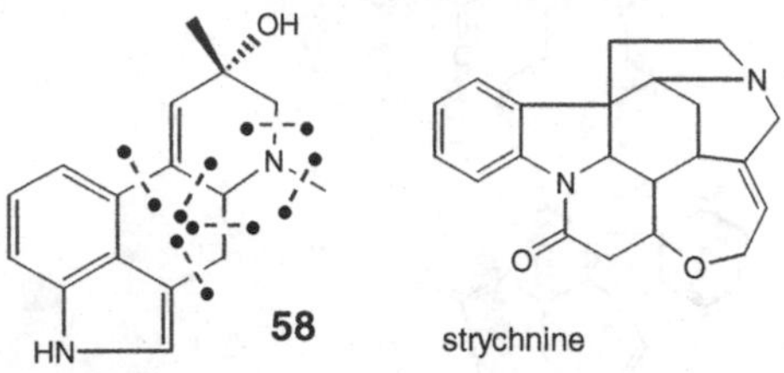

Scheme 8.5 Bond-set of the setoclavine synthesis in Scheme 8.4

At this point one could reach the nearly trivial conclusion that a synthesis is better, the shorter it is (the least number of bonds in the bond-set); the fewer the number of refunctionalization steps involved (oxidations, reductions); the fewer the number of protecting group steps; and the more it relies on reactions that form two or more skeletal bonds at a time. This accounts for the significant attention that is presently being given to tandem reactions [10] and reaction cascades.

It is a cardinal principle of synthesis that it evolves from simple starting materials at the beginning to a complex structure in the end. During a synthesis sequence the complexity of the intermediates increases in a discontinuous manner. The *increase in complexity*—more precisely, how rapid it is and at which point of the synthetic sequence it occurs—forms another important criterion to rank syntheses. The complexity of a target structure or of an intermediate is not, however, defined in a generally accepted manner. One could subscribe to the statement by Robinson [11] that strychnine (Scheme 8.5) is the most complex molecule known in relation to its size. When chemists speak of the complexity of a compound, they not only imply a complexity inherent in the structure, being topologically defined by the network of bonds. They also think about the difficulties of a synthesis, which may be a consequence of a lack of appropriate methods for synthesis [12]. Accordingly, adamantane (Scheme 8.6) was for a long time considered a rather complex target molecule. However, after a surprisingly simple synthesis of adamantane was realized by Schleyer [13], adamantane lost that status and is now a readily available commodity [14, 15]. Planning the synthesis of a compound like adamantane by a thermodynamics-driven equilibration process requires, however, completely different rules of synthesis planning than those presented in this text.

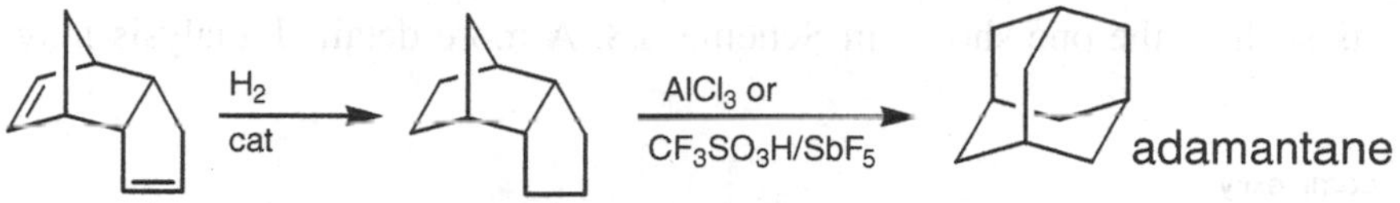

Scheme 8.6 Thermodynamics-driven formation of adamantane

How is the complexity of a structure related to synthesis planning? One should start with the following truism: reactions with simple molecules proceed readily and in high yields. Complex molecules are frequently "touchier" and tend to form side products even in simple looking transformations. While this is certainly not a natural law, most chemists would attribute considerable truth to this statement. The consequence for planning a synthesis is this: keep the number of steps low, once your intermediates have become increasingly complex. A rapid increase in complexity early in a synthesis sequence frequently faces low yields in the (many) operations that follow. Sequences involving overbred molecular skeletons, i.e., those with a complexity higher than the following intermediates or the target structure, can be justified only when they allow a substantial reduction in the overall number of steps. Based on the above considerations, a late (exponential) increase of complexity in a synthesis sequence is most desirable. A most striking example is given in Scheme 8.7.

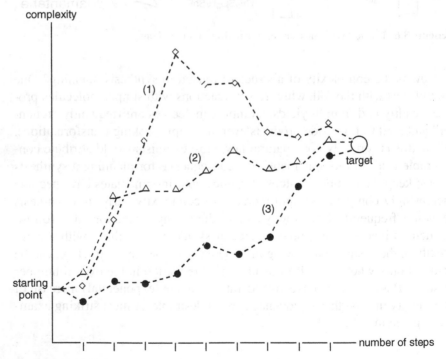

C. H. Heathcock [16]

Scheme 8.7 Rapid increase in complexity by a pentacyclization reaction

For analyses of synthesis sequences it will be of interest to quantify the complexity of both intermediates and final products. To this end, complexity indices have been developed which are based either on the topological complexity of the structures alone [12], or which, in addition, regard qualitatively the anticipated difficulties in a synthesis (intricacy indices) [17]. For a comparison of syntheses it is instructive to plot the "complexity" of the intermediates against the steps in the sequence. This results in plots such as the one shown in Scheme 8.8. A more detailed analysis may

Scheme 8.8 Complexity trendlines of the intermediates in several (hypothetical) synthetic sequences for a given target

incorporate the structural similarity between the intermediates and the target [18] as well.

The charts arrived at in this manner have been analyzed by Bertz [19] in a paper strongly recommended for careful reading [cf. also 17]. Bertz adds the complexities of all intermediates along a synthesis to give the overall complexity. One is not surprised to find that the higher the overall complexity, the lower is the overall yield! This mirrors common experience. Again, the bottom-line is: avoid refunctionalization steps once the complexity of the intermediates has become high.

Of the synthetic sequences depicted as complexity trendlines in Scheme 8.8, sequence (3) appears to be the most attractive one, as it features a low overall complexity with a late-stage, rapid increase in complexity. In sequence (2), the complexity rises much too quickly—affording high overall complexity. Finally, sequence (1) features intermediates with complexity exceeding that of the target; hence, it should be given the lowest rank. Such an analysis allows one to pinpoint deficiencies in synthetic sequences, and to determine at which point in a synthetic sequence improvements would be of the greatest consequence [17].

Such reasoning will favor syntheses that feature an exponential (= late) increase in complexity [20]. This can be influenced by the order in which the individual bonds in a bond-set are constructed. Here, the terms "linear" or "convergent" or "partially convergent" come into play. These terms were coined by Velluz as a consequence of his search for an optimal assembly of the steroid skeleton [21]. Convergent syntheses will afford a much better material balance than linear sequences, as soon as the yields in the individual steps fall below being quantitative. This point is illustrated in Schemes 8.9 and 8.10 for the coupling of eight building blocks (A through H) to yield a final target ABCDEFGH assuming an average yield of 80% for each coupling step.

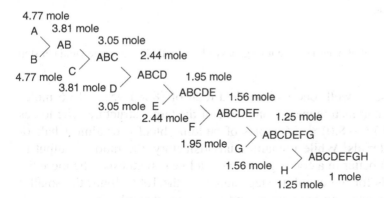

Scheme 8.9 Input of material for a linear synthesis with 80% average yield per step

If each and every coupling reaction proceeded quantitatively, only one mole of each of the starting materials would be needed to generate one mole of the target compound. If the yields for the coupling steps fall to about 80%, one is forced into a battle of materials: a total of 24 moles (instead of eight!) of starting materials is required to generate just one mole of the target compound. This means that 16 out of 24 moles (= to 2/3) of the material is lost in the form of side products or other losses—that is, waste that must be handled. Likewise, use of auxiliary reagents, neutralizing agents, and so forth is equally uneconomical.

In a fully convergent synthesis the situation is not perfect either, yet much more favorable (Scheme 8.10).

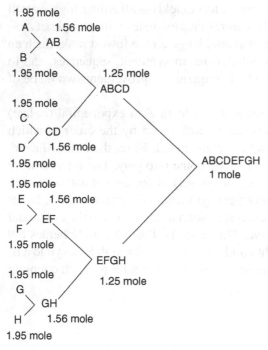

Scheme 8.10 Input of material for convergent synthesis with 80% average yield per step

In this case as well, one still needs a total of $(8 \times 1.95) = 15.6$ moles of starting materials to generate one mole of the target structure. The losses amount to $(15.6 - 8.0) = 7.6$ moles of building blocks, or almost half of the input materials. While remaining unsatisfactory, the ratio of output to input is much better in a convergent than in a linear synthesis. The more the average yields for the coupling steps approach the 100% limit, the smaller the difference gets between a linear and a convergent synthesis.

In this context the differences in "total yield" are frequently mentioned. "Total yield" refers to the yield of the final product ABCDEFGH calculated on the input of the initial first building block A. In a linear synthesis with average yields of 80% per step, the total yield for seven steps would amount to $0.8^7 = 21\%$. Since, in the hypothetical convergent synthesis, building block A would be carried through only three construction steps, the total yield would amount to $0.8^3 = 51\%$. Total yield is not a valid measure of the quality of a synthesis, because it measures the utilization of only the first building block, A. Utilization of a building block introduced later into the sequence will be higher (the yield based on that building block) as it is subjected to fewer overall yield-decreasing steps [22]. This is the basis for the advice to introduce "expensive" or otherwise "elaborate" building blocks as late as possible in a synthesis sequence!

Because of its historic importance for the development of the convergency principle, we summarize in Scheme 8.11 the estrone synthesis by Velluz [23].

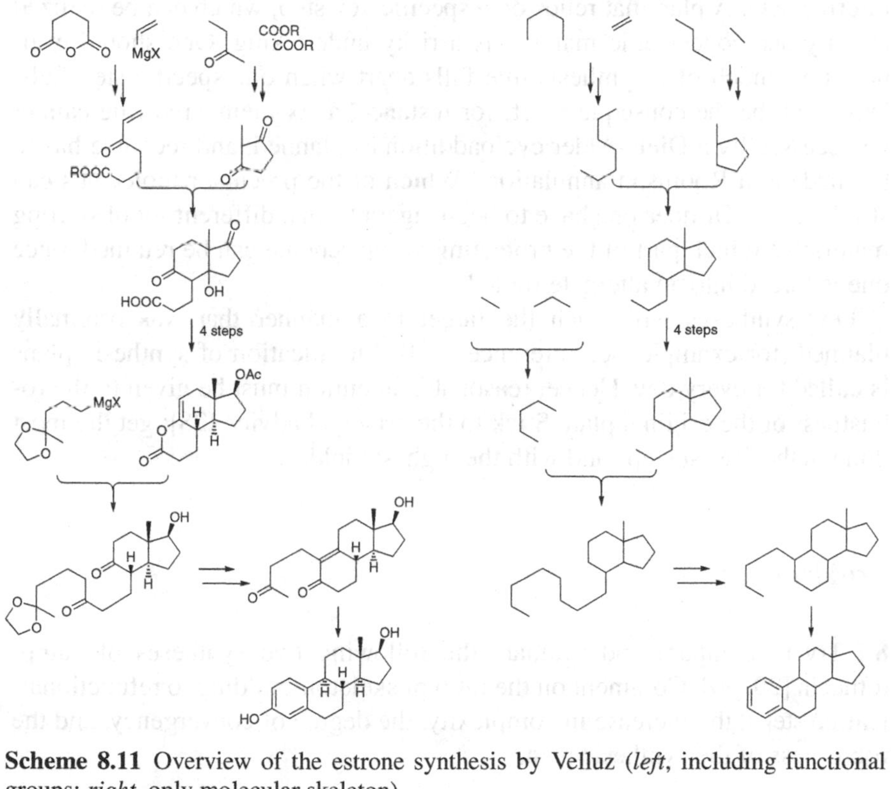

Scheme 8.11 Overview of the estrone synthesis by Velluz (*left*, including functional groups; *right*, only molecular skeleton)

In the left half of Scheme 8.11 the full sequence of the estrone synthesis is given. In the right half merely the construction plan (i.e., the skeletal atoms and connectivities) [22, 24] is shown, highlighting the (partial) convergency [25].

In summary, a good synthesis plan should be convergent, equivalent to a late-stage increase in complexity. The skeletal bond-forming steps should dominate in the step count. This means that refunctionalization steps and protecting group operations should be minimal, if they cannot be avoided entirely. In order to help meet these goals, consider the use of latent functionality.

Plans are subject to adverse events that may culminate in failure. This leads to the question: How *robust* is the plan to begin with? [26] One should check which and how many alternatives are available for each of the steps in a synthesis, in case the originally chosen variant cannot be employed. A robust synthesis plan will have at least one alternative for each step. Alternatives may include a change in the order in which bond-formations or refunctionalizations are to be carried out, or a longer sequence to reach a key intermediate. A plan that relies on a specific key step, which can be realized in only one conceivable manner, is a risky undertaking. One should evaluate how much of a synthesis tree falls apart when one specific step fails. What will be the consequence if, for instance, a six-membered ring cannot be accessed by a Diels-Alder cycloaddition as planned, and recourse has to be made to a Robinson-annulation? Which of the precursor molecules can still be used? Or does one have to begin again from a different set of starting materials? Which part of the protecting group scheme can be retained, once one is forced into an alternate route?

Few syntheses will reach the target in a manner that was originally planned (for example, see reference [27]). Modification of synthesis plans is called for every day. Hence, reasonable attention must be given to the robustness of the original plan. Stick to the principal advice [28]: get the most done in the fewest steps and with the highest yield.

Problems

8.1 Try to compare and evaluate the following two syntheses of camptothecin [29, 30]. Comment on the ratio of skeleton-building to refunctionalization steps, the increase in complexity, the degree of convergency, and the robustness of the synthesis plan.

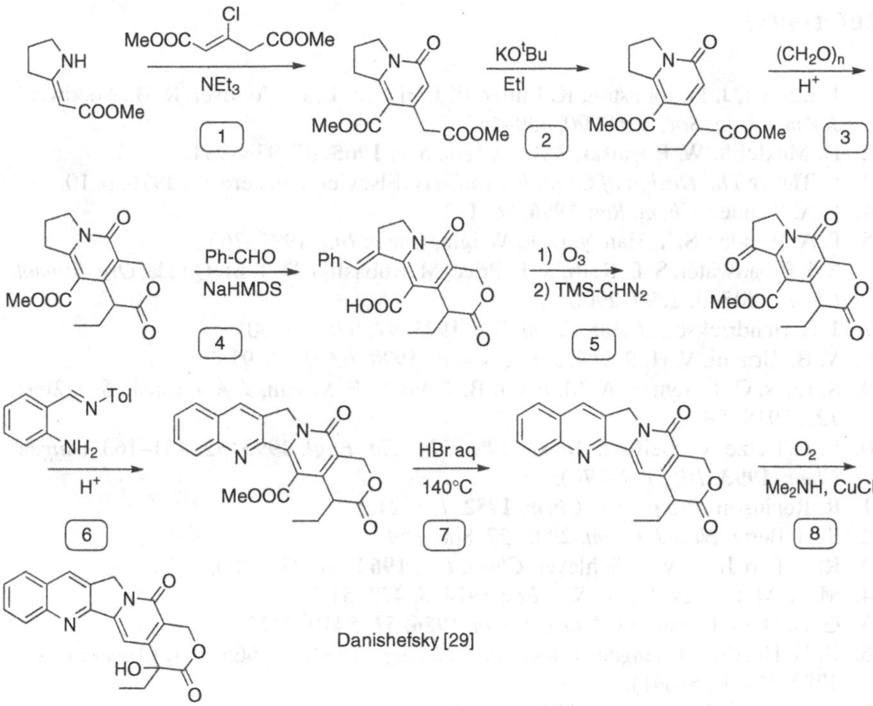

Scheme 8.12 Danishefsky's camptothecin synthesis

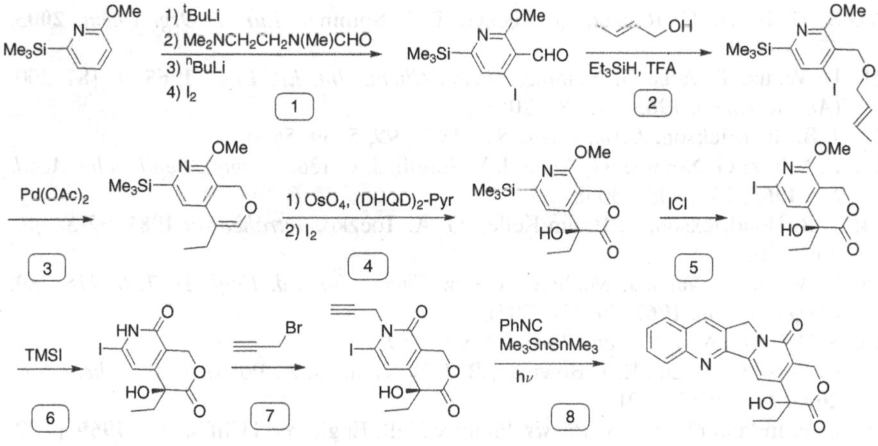

Scheme 8.13 Curran's camptothecin synthesis

References

1. J. J. Korst, J. D. Johnston, K. Butler, E. J. Bianco, L. H. Conover, R. B. Woodward, *J. Am. Chem. Soc.* **1968**, *90*, 439–457.
2. H. Muxfeldt, W. Rogalski, *J. Am. Chem. Soc.* **1965**, *87*, 933–934.
3. S. Turner *The Design of Organic Synthesis*, Elsevier, Amsterdam, **1976**, p. 10.
4. P. A. Wender, *Chem. Rev.* **1996**, *96*, 1–2.
5. P. A. Wender, S. T. Handy, D. L. Wright, *Chem. Ind.* **1997**, 765.
6. S. J. Broadwater, S. L. Roth, K. E. Price, M. Kobaslija, D. T. McQuade, *Org. Biomol. Chem.* **2005**, *3*, 2899–2906.
7. J. B. Hendrickson, *J. Am. Chem. Soc.* **1975**, *97*, 5784–5800.
8. V. B. Birman, V. H. Rawal, *J. Org. Chem.* **1998**, *63*, 9146–9147.
9. S. Liras, C. L. Lynch, A. M. Fryer, B. T. Vu, S. F. Martin, *J. Am. Chem. Soc.* **2001**, *123*, 5918–5924.
10. L. F. Tietze, U. Beifuss, *Angew. Chem. Int. Ed. Engl.* **1993**. *32*, 131–163. (*Angew. Chem.* **1993**, *105*, 137–170).
11. R. Robinson, *Progr. Org. Chem.* **1952**, *1*, 1–21.
12. S. H. Bertz, *New J. Chem.* **2003**, *27*, 860–869.
13. R. C. Fort Jr., P. v. R. Schleyer, *Chem. Rev.* **1964**, *64*, 277–300.
14. M. A. McKervey, *Chem. Soc. Rev.* **1974**, *3*, 479–512.
15. G. A. Olah, D. Farooq, *J. Org. Chem.* **1986**, *51*, 5410–5413.
16. C. H. Heathcock, *Angew. Chem., Int. Ed. Engl.* **1992**, *31*, 665–681. (*Angew. Chem.* **1992**, *104*, 675–691).
17. P. L. Fuchs, *Tetrahedron* **2001**, *57*, 6855–6875.
18. M. Chanon, R. Barone, C. Baralotto, M. Julliard, J. B. Hendrickson, *Synthesis* **1998**, 1559–1583.
19. S. H. Bertz, *J. Am. Chem. Soc.* **1982**, *104*, 5801–5803.
20. S. H. Bertz, C. Rücker, G. Rücker, T. J. Sommer, *Eur. J. Org. Chem.* **2003**, 4737–4740.
21. L. Velluz, J. Valls, G. Nominé, *Angew. Chem., Int. Ed. Engl.* **1965**, *4*, 181–200. (*Angew. Chem.* **1965**, *77*, 185–205).
22. J. B. Hendrickson, *J. Am. Chem. Soc.* **1977**, *99*, 5439–5450.
23. L. Velluz, G. Nominé, G. Amiard, V. Torelli, J. Cérêde, *Compt. Rend. hebd. Acad. Sci.* **1963**, *257*, 3086–3088.
24. J. B. Hendrickson, E. Braun-Keller, G. A. Toczko, *Tetrahedron* **1981**, *37 Suppl.*, 359–370.
25. L. Velluz, J. Valls, J. Mathieu, *Angew. Chem., Int. Ed. Engl.* **1967**, *6*, 778–789. (*Angew. Chem.* **1967**, *79*, 774–785).
26. S. H. Bertz, *New J. Chem.* **2003**, *27*, 870–879.
27. R. Shen, C. T. Lin, E. J. Bowman, B. J. Bowman, J. A. Porco, *J. Am. Chem. Soc.* **2003**, *125*, 7889–7901.
28. R. E. Ireland *Organic Synthesis*, Prentice Hall, Englewood Cliffs, N.J., **1969**, p. 29.
29. W. Shen, C. A. Coburn, W. G. Bornmann, S. J. Danishefsky, *J. Org. Chem.* **1993**, *58*, 611–617.
30. D. P. Curran, S.-B. Ko, H. Josien, *Angew. Chem., Int. Ed. Engl.* **1995**, *34*, 2683–2684. (*Angew. Chem.* **1995**, *107*, 2948–2950).

Chapter 9
Computer-Aided Synthesis Planning

Abstract Computational approaches exist to support synthesis planning. Knowledge-based programs utilize databases of known reactions, whereas logic-based programs analyze the topology of the target structure to suggest optimal—not necessarily precedented—synthesis routes.

Callystatin A (Scheme 9.1) serves as an example of a target molecule of moderate complexity. It has 28 skeletal bonds and five bonds between the backbone and heteroatoms. The number of possible bond-sets and construction sets is so large, that it is impossible for a single chemist to generate and rank them all.

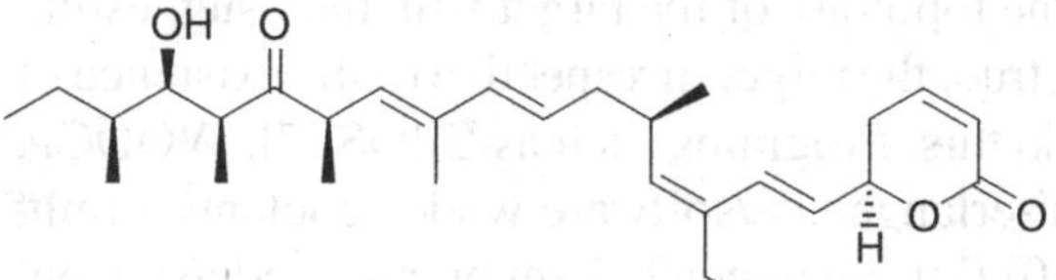

Scheme 9.1 Structure of callystatin A

When a chemist has arrived at ten projected synthesis plans for a molecule of this size, and when one plan by and large meets the criteria set by the chemist, he or she will no longer be willing to invest more time in planning. One must accept that several plans may have been overlooked that would have been significantly superior. A systematic search of the possibilities remains undone simply because of the size of the problem. However, the systematic processing of large amounts of information is within the domain of computers; this prompted an analysis of the pros and cons of computer-aided synthesis planning [1, 2].

Historically, this analysis set the stage for systematizing the intellectual aspects of synthesis planning and for uncovering any logical pattern by which the process of synthesis planning might be described. This in turn

R.W. Hoffmann, *Elements of Synthesis Planning*,
DOI 10.1007/978-3-540-79220-8_9, © Springer-Verlag Berlin Heidelberg 2009

led [3] to Corey's *Logic of Chemical Synthesis* [4] which earned him the Nobel Prize in 1990. In preparation for computer-aided synthesis planning, three tasks had to be mastered:

(i) the mapping of chemical structures in a computer-readable manner,
(ii) the mapping of chemical reactions in a computer-readable manner, and
(iii) the establishment of criteria for ordering and ranking synthesis plans in a computer-readable manner

Once tasks (i) and (ii) had been accomplished, a computer would be able to retrosynthetically disconnect a target structure and list all theoretically possible routes to construct the target. An astronomically high number of synthesis trees was generated. At this stage it became essential to limit these to only the most meaningful solutions. This pruning of the syntheses trees could be done in an interactive fashion. In this case it is the chemist who would make (consequential) wrong decisions. Alternatively, one could leave this task to the computer, which sorts the possibilities according to point (iii) above. The computer then presents the results starting with the proposals that earned the highest scores. This seems acceptable, once you are confident enough that the computer is able to handle the ranking criteria in a proper manner. These implications indicate that there will be no single "ideal" computer program for the planning of syntheses.

Starting about 1970, several programs were developed worldwide for the planning of syntheses [5]. They have significant differences [6]. There are those programs that focus on the topology of the target structure, suggesting how one could and should construct the target, irrespective of the existence of synthetic methods that could do this. Programs such as EROS [7], WODCA [8, 9] (http://www2.chemie.uni-erlangen.de/software/wodca/contents. html# overview_c), and SYNGEN [10] (http://syngen2.chem.brandeis.edu/syngen. html) fall within this category. These "logic-based" programs address the process of synthesis planning from a fundamental aspect that offers opportunities to delineate completely new chemical transformations. Other programs such as LHASA (http://lhasa.harvard.edu/) or (http://cheminf.cmbi.ru. nl/cheminf/lhasa/doc/lhasa191.pdf) are built on a huge database of reactions, guaranteeing that the proposals made will have a welcome precedence in reliable chemistry. The LHASA program does not only know simple transformations, but evaluates the possibilities to implement certain key transformations such as the Diels-Alder cycloaddition into the synthesis plan. It suggests the introduction of auxiliary functionality (FGA) to enable the implementation of these reaction schemes. In LHASA, the synthesis trees are pruned both in an interactive fashion and by the program itself in order to focus on attractive proposals. For an example see in Scheme 9.2 an analysis for the synthesis of the sesquiterpene valeranone. E. J. Corey commented [3]: "The suggestion by LHASA of such nonobvious pathways is both stimulating and valuable to a chemist. The field of computer-assisted synthetic analysis is fascinating in its own right, and surely one of the most interesting problems in the area of machine intelligence. Because of the enormous memory and speed of modern machines and the probability of continuing advances, it seems clear that computers can play an important role in synthetic design."

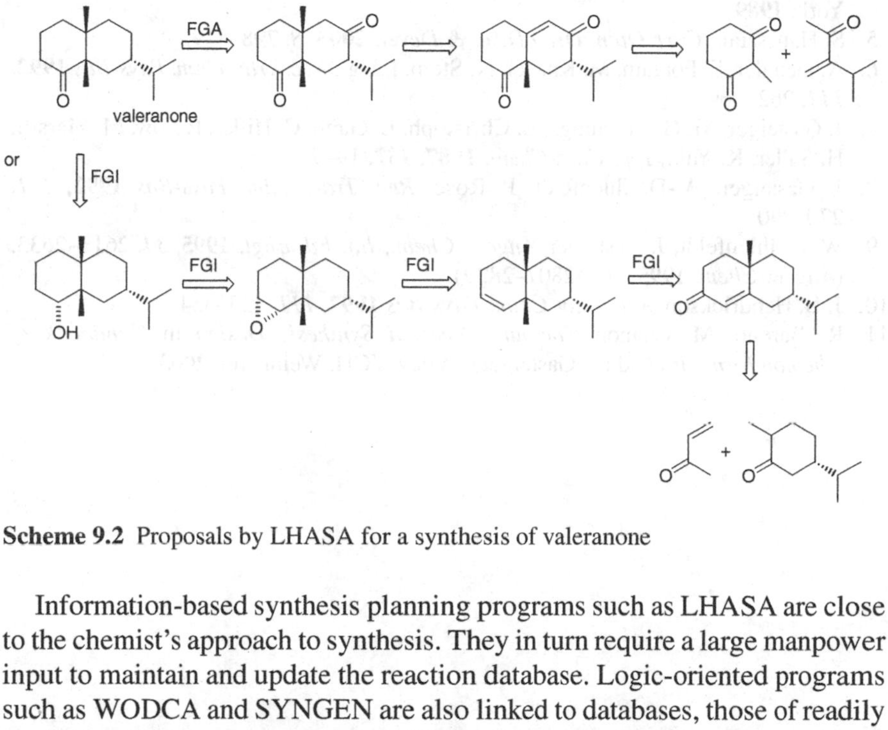

Scheme 9.2 Proposals by LHASA for a synthesis of valeranone

Information-based synthesis planning programs such as LHASA are close to the chemist's approach to synthesis. They in turn require a large manpower input to maintain and update the reaction database. Logic-oriented programs such as WODCA and SYNGEN are also linked to databases, those of readily available simple starting materials. These programs check the structural similarity of the target or intermediates to potential starting materials in order to suggest the shortest route to the target.

Overall, the acceptance of such synthesis planning programs up to now has remained rather low, considering all the progress that has been made in the development of such programs [1, 11]. There remains the task to improve the interface of these programs to make them really attractive in the future and to maximize their potentials.

References

1. M. H. Todd, *Chem. Soc. Rev.* **2005**, *34*, 247–266.
2. M. Sitzmann, M. Pförtner *Computer-Assisted Synthesis Design* in *Chemoinformatics* (Eds.: J. Gasteiger, T. Engel), Wiley-VCH, Weinheim, **2003**.
3. E. J. Corey, *Angew. Chem., Int. Ed. Engl.* **1991**, *30*, 455–465. (*Angew. Chem.* **1991**, *103*, 469–479).
4. E. J. Corey, X.-M. Cheng, *The Logic of Chemical Synthesis*, J. Wiley & Sons, New York, **1989**.
5. S. Hanessian, *Curr. Opin. Drug Disc. & Devel.* **2005**, *8*, 798–819.
6. A. Dengler, E. Fontain, M. Knauer, N. Stein, I. Ugi, *Rec. Trav. chim. Pays-Bas* **1992**, *111*, 262–269.
7. J. Gasteiger, M. G. Hutchings, B. Christoph, L. Gann, C. Hiller, P. Löw, M. Marsili, H. Saller, K. Yuki, *Top. Curr. Chem.* **1987**, *137*, 19–73.
8. J. Gasteiger, W.-D. Ihlenfeldt, P. Röse, *Rec. Trav. chim. Pays-Bas* **1992**, *111*, 270–290.
9. W.-D. Ihlenfeldt, J. Gasteiger, *Angew. Chem., Int. Ed. Engl.* **1995**, *34*, 2613–2633. (*Angew. Chem.* **1995**, *107*, 2807–2829).
10. J. B. Hendrickson, *Rec. Trav. Chim. Pays-Bas* **1992**, *111*, 323–334.
11. R. Barone, M. Chanon *Computer-Assisted Synthesis Design* in *Handbook of Chemoinformatics* (Ed.: J. Gasteiger), Wiley-VCH, Weinheim, **2003**.

Chapter 10
Stereogenic Centers and Planning of Syntheses

Abstract Isolated stereogenic centers of chiral target structures and first stereogenic centers of sequences of neighboring stereogenic centers are generated by asymmetric synthesis, or by resolution of a racemate, or acquired from the chiral pool. Sequences of neighboring stereogenic centers are generated from an initial one by methods of diastereoselective synthesis. They are frequently perceived as patterns, for the synthesis of which special methods are established.

Forty years ago, stereogenic centers in a target structure were cause for serious headaches in synthesis planning. Since then, the methodology of stereoselective synthesis has been advanced to the point that aspects of stereogenic centers, while still important, are no longer the major hurdle in planning syntheses. For this reason, the relevant discussion was placed last (but not least) in this treatise.

When a target structure has more than one stereogenic center, start by assessing the distance from one to the other. When the stereogenic centers are adjacent or at a distance not greater than 1,4, consider constructing these stereogenic centers either simultaneously or sequentially, the second one by asymmetric induction from the first. In case the distance between the stereogenic centers is greater than 1,4, these centers should be treated independently. In any case, the main focus is on how to generate the first stereogenic center in a projected synthesis [1]. The available possibilities include:

- stereoselective synthesis, e.g., by a stereogenic skeletal bond-forming reaction;
- asymmetric synthesis, e.g., by a stereogenic refunctionalization reaction, including enzymatic approaches;
- synthesis of a racemate, followed by resolution;
- incorporation of a chiral building block from the chiral pool.

An initial feeling for these alternatives can be gained by discussion of the synthesis of sulcatol (Scheme 10.1), a pheromone of a noxious insect. The

R.W. Hoffmann, *Elements of Synthesis Planning*,
DOI 10.1007/978-3-540-79220-8_10, © Springer-Verlag Berlin Heidelberg 2009

species *Gnathotrichus retusus* is excited by the pure (*S*)-enantiomer. This action is inhibited by the other enantiomer. The related species *Gnathotrichus sulcatus* does not react to either pure enantiomer, but to a 65/35 mixture of (*S*)- and (*R*)-sulcatol [2]. The synthesis is thus challenged to provide routes to each pure enantiomer of the compound.

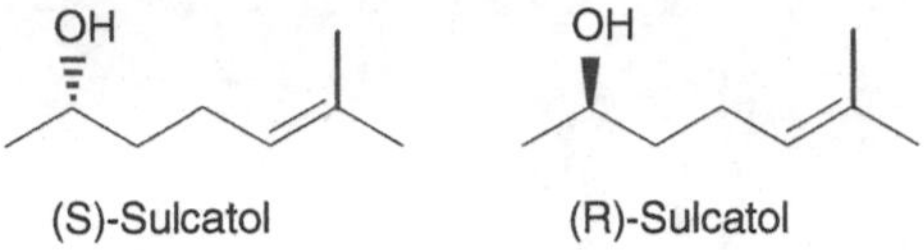

(S)-Sulcatol   (R)-Sulcatol

Scheme 10.1 The two enantiomeric forms of sulcatol

When considering the formation of a stereogenic center by synthesis, one should evaluate the chances of forming each of the four bonds at the stereogenic center by a stereoselective bond formation (Scheme 10.2).

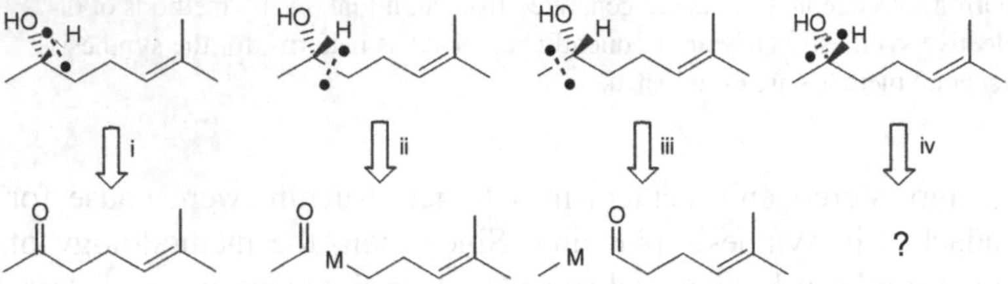

Scheme 10.2 Retrosynthetic considerations regarding the stereogenic center in sulcatol

Sulcatol, a chiral alcohol, could be approached using an enantioselective reduction of a ketone (bond-set (i) in Scheme 10.2) applying, e.g., the Corey–Bakshi–Shibata method [3]. In the present case, reduction with various strains of yeasts appears equally attractive [2] According to Hendrickson's definition of an ideal synthesis, stereogenic centers should be established during skeletal bond-forming reactions (cf. bond-sets (ii) and (iii) in Scheme 10.2). This suggests adding, e.g., allyl-di-isopinocampheyl-borane to acetaldehyde to give the homoallylic alcohol **59** (Scheme 10.3) [4], that could be converted into sulcatol by a few further steps.

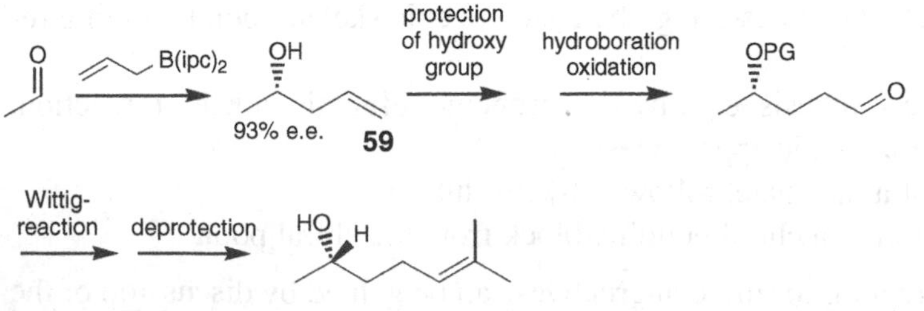

Scheme 10.3 Asymmetric allyboration to generate the stereogenic center of sulcatol

Bond-set (iii) in Scheme 10.2 for the synthesis of sulcatol could be realized with an umpoled synthon, applying Hoppe's carbamate procedure [5] (Scheme 10.4).

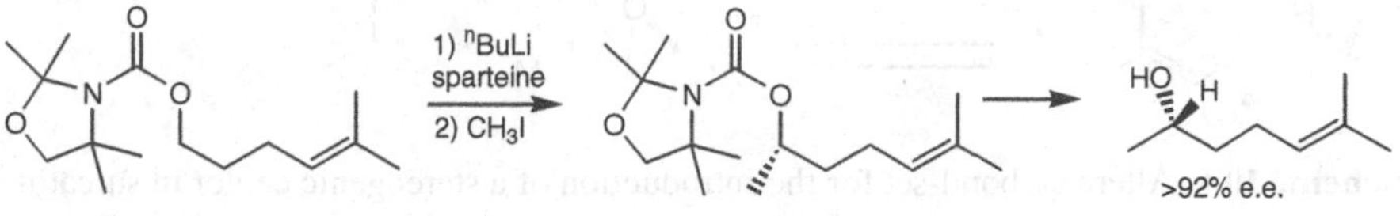

Scheme 10.4 Asymmetric deprotonation to generate the stereogenic center of sulcatol

This shows that in many cases the generation of a stereogenic center requires additional steps, be it in the substrate or the reagent, such as the attachment of the carbamate moiety and its ultimate removal, in the example in Scheme 10.4. The covalent attachment of the substrate to chiral or achiral auxiliaries decreases synthetic efficiency, as does the introduction and removal of a protecting group. Hence one should evaluate whether the required stereogenic center could not more profitably be garnered from the chiral pool. This suggests that one look for chiral building blocks that contain a doubly bound heteroatom at C-1 and a secondary alcohol function at C-4. Perusal of the compilation by Scott [6] produced the potential precursors listed in Scheme 10.5.

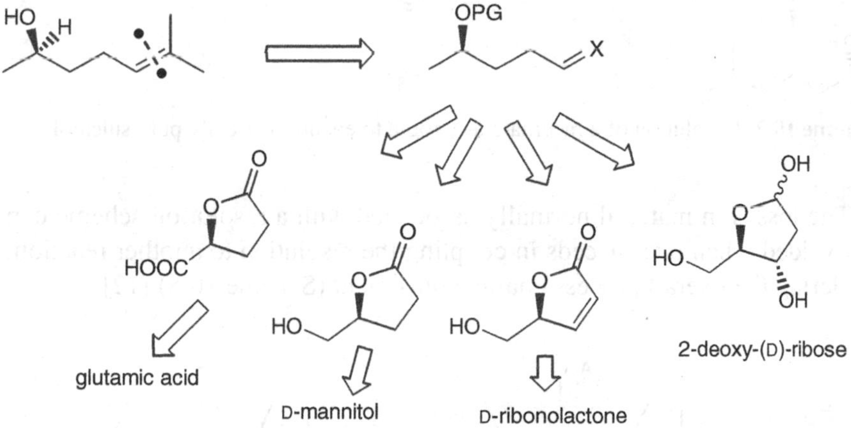

Scheme 10.5 Potential chiral precursors for a synthesis of sulcatol

In due course, syntheses of sulcatol have been reported that were based on glutamic acid [7] or on 2-deoxy-ribose [8]. They required, though, that five of the seven steps before the concluding Wittig-reaction be refunctionalizations. By today's standard this is unacceptable for the introduction of just

a single stereogenic center. Perhaps an alternate bond-set would fare better
(Scheme 10.6)?

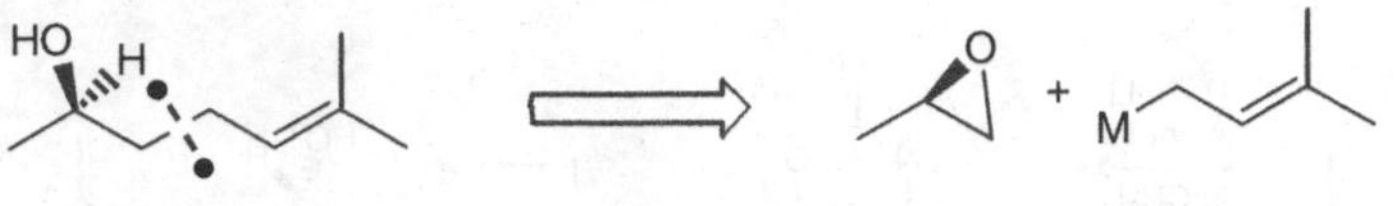

Scheme 10.6 Alternate bond-set for the introduction of a stereogenic center in sulcatol

Enantiomerically pure propeneoxide can be obtained in three steps from
lactic acid. Its reaction with prenyl cuprate directly yields sulcatol [9, 10].
Even with such a convincing route at hand, one should not fail to evaluate routes via a racemate. A classical resolution via the formation of a
hemiphthalate and crystallization of its brucine salt appears circumstantial.
Yet kinetic resolution using enzymes, e.g., lipases, appears more attractive
(Scheme 10.7) [11].

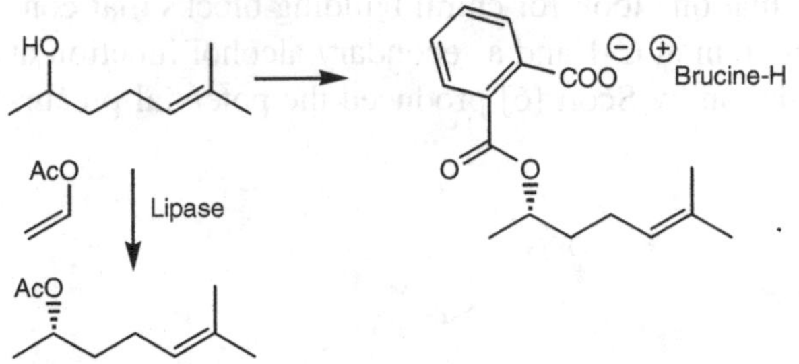

Scheme 10.7 Resolution of a racemate as a route to enantiomerically pure sulcatol

The losses in material normally associated with a resolution scheme can
be avoided when one succeds in coupling the resolution to another reaction,
rendering the overall process enantioconvergent (Scheme 10.8) [12].

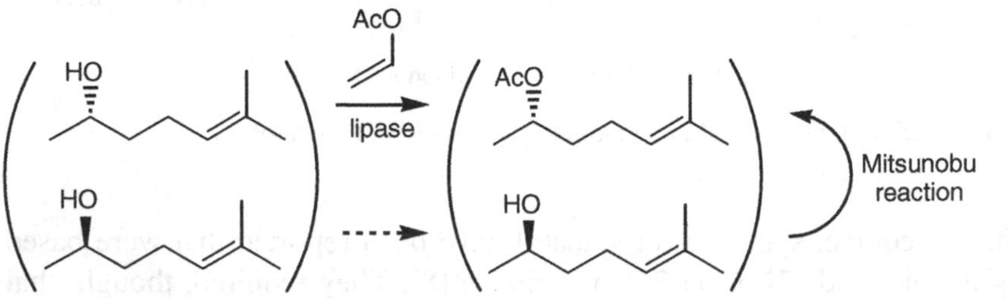

Scheme 10.8 Enantioconvergent resolution as a route to enantiomerically pure sulcatol

As a consequence, once a racemate is readily available, a resolution approach to obtain enantiomerically pure material is as valid today as in earlier times. In a multistep synthesis of targets having several stereogenic centers, a resolution approach should be limited to those cases where it can be applied early on in the synthesis, i.e., to obtain the first stereogenic center.

The example of sulcatol is representative of a comparatively simple starting building block with just one stereogenic center, a secondary alcohol. A similar example with a methyl branch as the stereogenic center is discussed next (Scheme 10.9). Start the retrosynthetic disconnections by considering a cut at every one of the four bonds at the stereogenic center. In order to generate a stereogenic center in the forward direction, the precursor must be prochiral. This is most readily achieved with an sp^2-hybridized carbon atom that may be part of a carbon-carbon double bond. Following this reasoning, a chiral methyl branch should be retrosynthetically connected to a set of prochiral olefins.

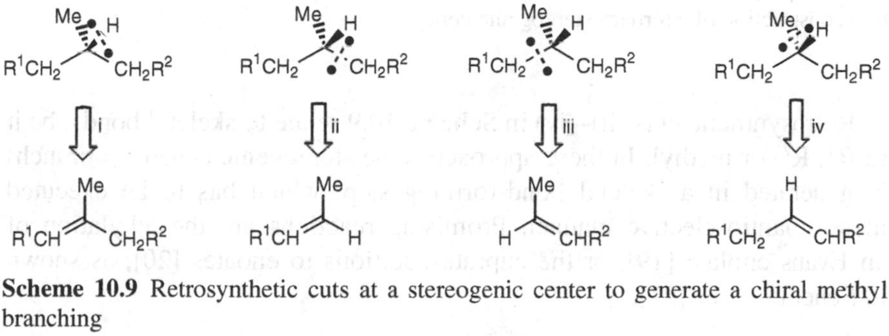

Scheme 10.9 Retrosynthetic cuts at a stereogenic center to generate a chiral methyl branching

Retrosynthesis (i) in Scheme 10.9 identifies stereoselective formation of a carbon-hydrogen bond as a route to the chiral methyl branch. This could be accomplished, for instance, by asymmetric protonation of a prochiral enolate, by asymmetric hydride transfer to a prochiral enoate (Scheme 10.10) [13], by transfer of a hydrogen atom to a prochiral carbon-centered radical, or by hydrogenation of a prochiral carbon-carbon double bond. The latter might succeed with the aid of chiral catalysts [14, 15]. Otherwise, one has to resort to the attachment of chiral auxiliaries [16, 17].

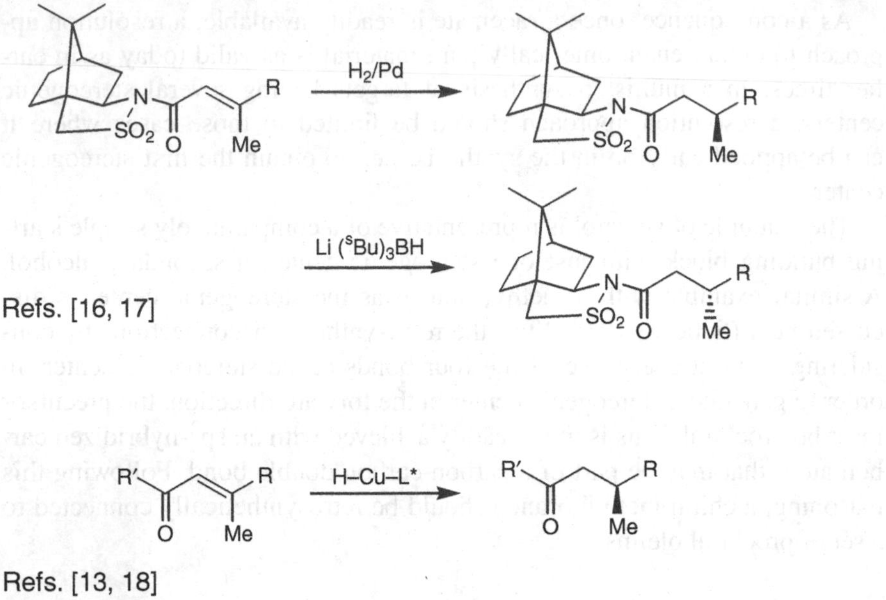

Refs. [16, 17]

Refs. [13, 18]

Scheme 10.10 Asymmetric reduction of unsaturated carbonyl systems for enantioselective construction of a tertiary stereogenic center

Retrosynthetic cuts (ii)–(iv) in Scheme 10.9 relate to skeletal bonds, be it to R^1, R^2, or methyl. In these approaches the stereogenic center (= branch) is generated in a skeletal bond-forming step, which has to be executed in an enantioselective manner. Promising reactions are the alkylation of an Evans enolate [19], or the cuprate additions to enoates [20], as shown in Scheme 10.11 [21].

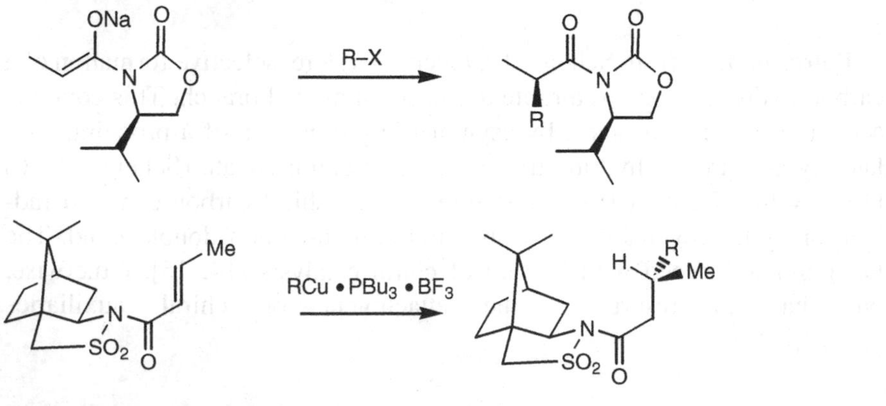

Scheme 10.11 Skeleton-forming reactions for the enantioselective formation of a tertiary stereogenic center

Compared to the multitude of methods to generate chiral secondary alcohols, the number of methods available to generate chiral methyl branches
appears rather limited. It thus becomes more attractive to look for methods by which the easily attained stereogenic center "secondary alcohol"
may be converted into the harder to obtain stereogenic center "methyl
branch." This puts the spotlight on sigmatropic rearrangements [22], as
well as on metal-mediated substitutions of the S_N2'-type (Scheme 10.12)
[23, 24].

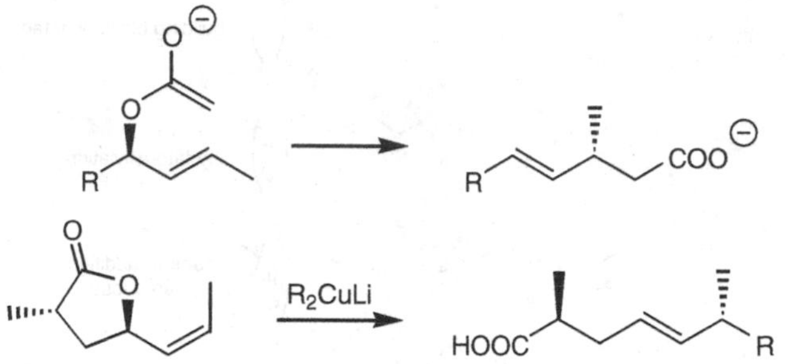

Scheme 10.12 Skeletal bond-forming rearrangements for the enantioselective generation of a tertiary stereogenic center

Once the possibilities of generating a stereogenic methyl branch by asymmetric synthesis have been explored, look into the possibilities of deriving
such an entity from the chiral pool. Natural products that could provide such
an entity include citronellal, which is available in both enantiomeric forms.
Other suitable members of the chiral pool are methyl 3-hydroxy-isobutyrate,
and 3-methylglutaric acid monoester, both derived from resolution of race-
mates [25] (Scheme 10.13).

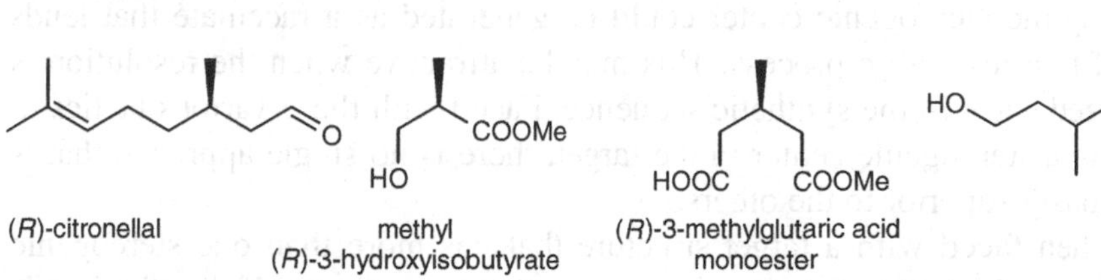

(*R*)-citronellal

methyl
(*R*)-3-hydroxyisobutyrate

(*R*)-3-methylglutaric acid
monoester

60

Scheme 10.13 Chiral precursor molecules with a tertiary methyl-bearing stereogenic
center

Resolution of racemic compounds carrying a methyl branch can be envisioned, when functionalities, such as a hydroxy group in compound **60** (Scheme 10.13), are in close proximity to the methyl branch to facilitate a lipase-catalyzed kinetic resolution [26].

The various approaches presented above for generating a stereogenic center at a methyl branch are mirrored in the bond-sets of syntheses of another insect pheromone shown in Scheme 10.14 [21, 27, 28].

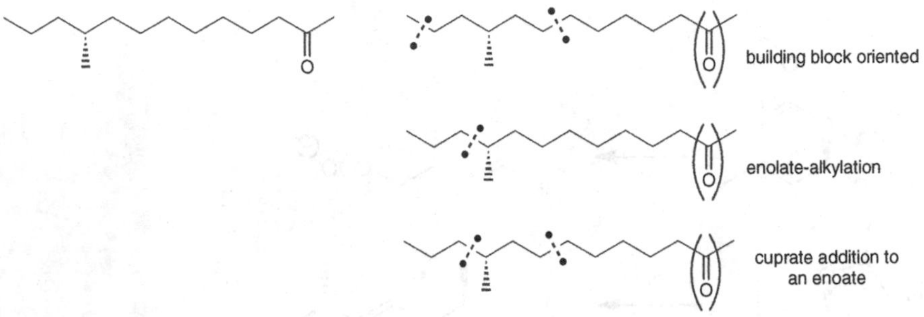

building block oriented

enolate-alkylation

cuprate addition to an enoate

Scheme 10.14 Bond-sets used to generate a tertiary stereogenic center in an enantioselective manner

The two examples, sulcatol and methyl-branched skeleton, illustrate how a stereogenic center influences actions in synthesis planning. Given the present status of synthetic methodology of enantioselective synthesis, the prime choice is to generate a stereogenic center by asymmetric synthesis and to modify the bond-set accordingly. A building block oriented approach, in which the stereogenic center is derived from precursors from the chiral pool, is in many cases less attractive because it requires a series of nonproductive steps to adjust the compound from the chiral pool to that actually needed for incorporation into the target. Moreover, check whether an intermediate containing the stereogenic center could be generated as a racemate that lends itself to a resolution process. This may be attractive when the resolution is realized early in the synthetic sequence. Faced with these various tactics to secure a stereogenic center in the target, there is no single approach that is obviously superior to the others.

When faced with a target structure that has more than one stereogenic center, evaluate the distance between stereogenic centers. If the distance is greater than 1,4, then one tends to consider them as independent entities. If the distance is 1,4 or smaller, one should try to establish both of them in a combined operation. Scheme 10.15 gives examples with a 1,3-distance of stereogenic centers.

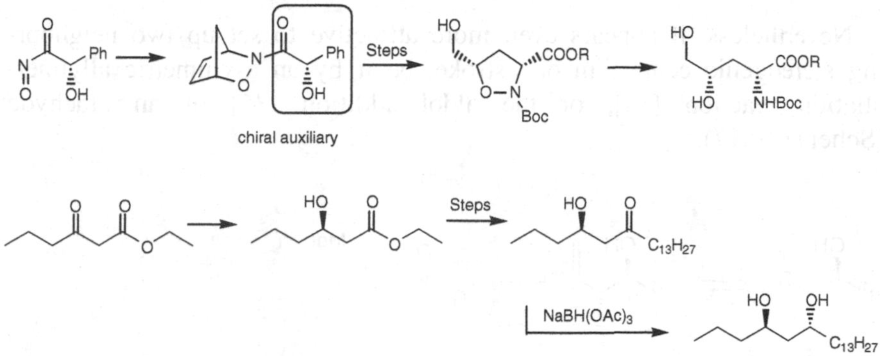

Scheme 10.15 Simultaneous or sequential establishment of two stereogenic centers in a 1,3-relationship

In the first [29] of the examples in Scheme 10.15, a hetero-Diels-Alder cycloaddition is realized under the asymmetric induction from a chiral auxiliary. In this cycloaddition two stereogenic centers in the five-membered heterocycle are set, yet a number of steps are required to remove the chiral auxiliary and to refunctionalize the intermediate to reach the final product. In the second example [30], a first stereogenic center is generated by a catalytic asymmetric reduction of a ketone function. This center then serves as the source of chiral information for the generation of the second stereogenic center by 1,3-asymmetric induction.

When two stereogenic centers are present in a 1,2-distance, the broad body of knowledge regarding 1,2-asymmetric induction [31, 32, 33] suggests sequential access to these two stereogenic centers (Scheme 10.16).

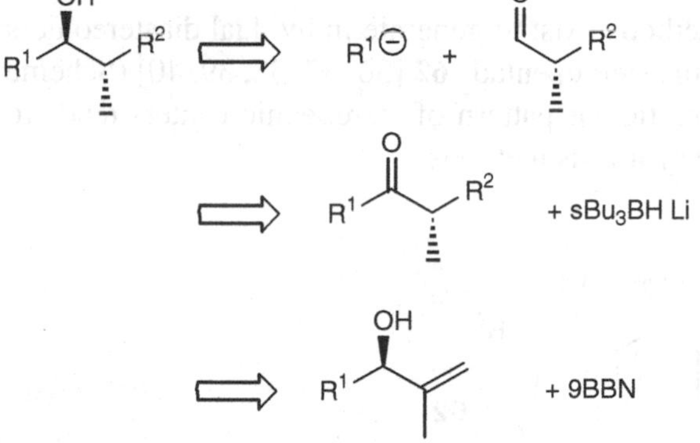

Scheme 10.16 Sequential construction by asymmetric induction of two stereogenic centers bearing a 1,2-relationship

Nevertheless, it appears even more attractive to set up two neighboring stereogenic centers in one stroke, be it by an asymmetric allylmetallation reaction [34], or the aldol addition [35] of an aldehyde (Scheme 10.17).

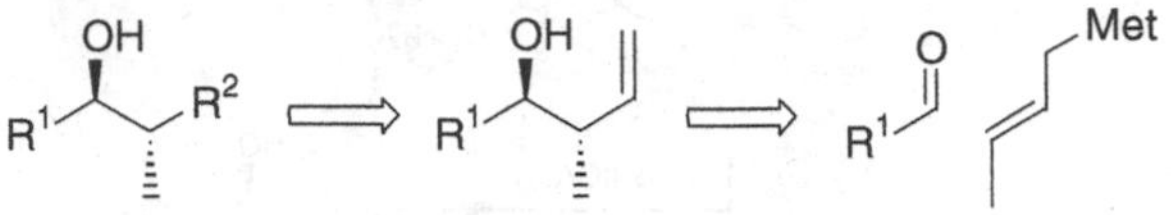

Scheme 10.17 Simultaneous construction of two stereogenic centers in a 1,2-relationship by enantioselective allylmetallation

Chemists tend to look at sequences of neighboring stereogenic centers as a sort of pattern recognition. These patterns are then associated with established methods to create such a pattern. This is illustrated in Scheme 10.18 concerning chiral 1,2-diols.

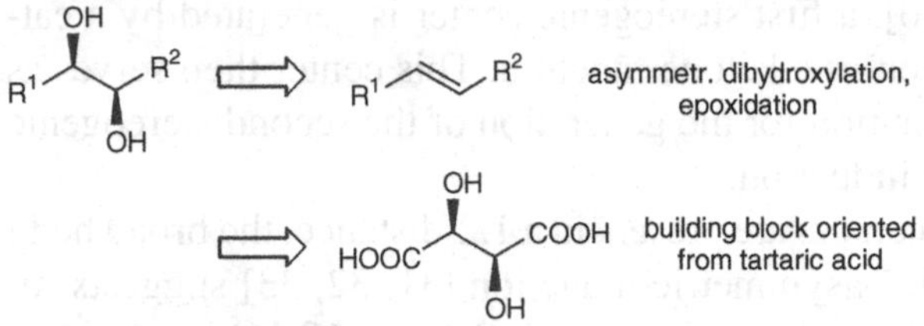

Scheme 10.18 Established retrosyntheses for chiral 1,2-diols

Special collections of methods exist to generate individual diastereomers of the stereotriads **61** or of the stereopentads **62** [36, 37, 38, 39, 40] (Scheme 10.19). In these cases the particular pattern of stereogenic centers tends to determine the choice of the synthesis methods.

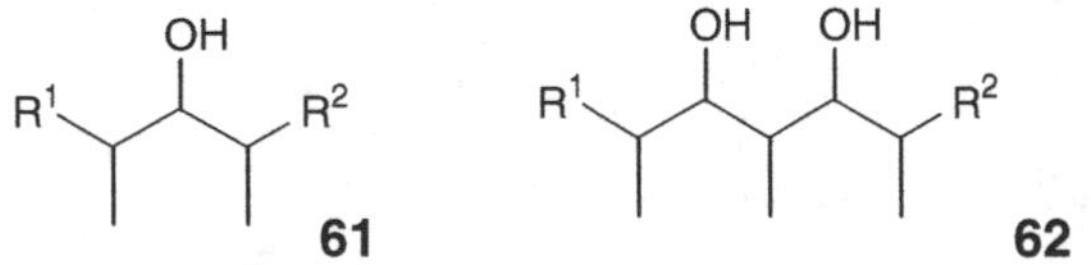

Scheme 10.19 Stereotriads and stereopentads of neighboring stereogenic centers

Along these lines, one gradually shifts to method-oriented retrosynthesis. An example of such a strategy is given by a route to erythronolide A (Scheme 10.20), during which two key reactions, the asymmetric crotylboration of aldehydes and the Sharpless epoxidation, feature prominently and dictate the plan for the synthesis [41, 42].

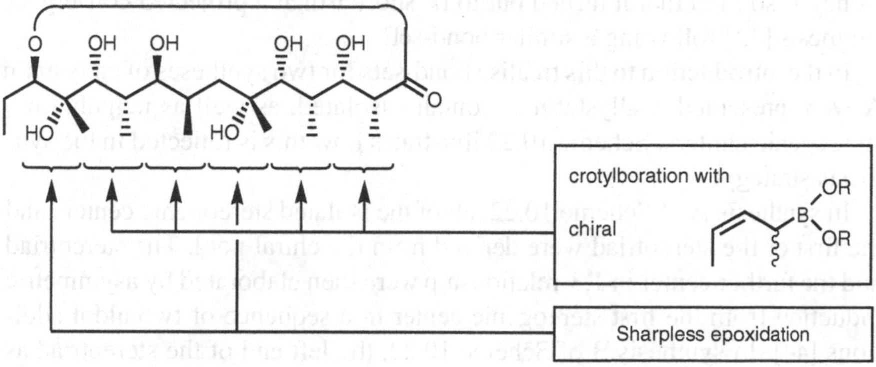

Scheme 10.20 Method-oriented construction of the neighboring stereogenic centers in erythronolide A, featuring reagent control of stereoselectivity

The individual steps of the stereoselective crotylboration are detailed in Scheme 10.21, which demonstrates how establishment of the stereogenic centers is linked to the skeletal bond-forming reactions.

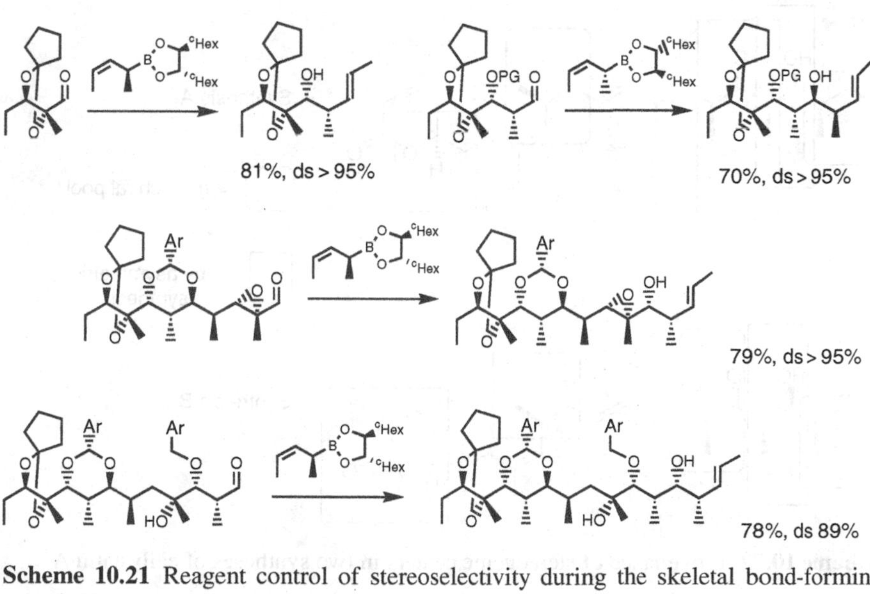

Scheme 10.21 Reagent control of stereoselectivity during the skeletal bond-forming steps in the synthesis of erythronolide A

Scheme 10.21 features just those steps in which the stereogenic centers and the skeleton are being formed. Not shown are the refunctionalization steps which were necessary in order to convert the product of one skeletal bond-forming reaction into the starting point for the next. One should note that eight of the eleven stereogenic centers were established during skeletal bond-forming reactions. This route to erythronolide A is linear, yet its efficiency is so high that it turned out to be shorter than a projected convergent synthesis [43] following a similar bond-set!

In the introduction to this treatise bond-sets for two syntheses of callystatin A were presented. Callystatin A contains isolated, as well as neighboring, stereogenic centers. Scheme 10.22 illustrates how this is reflected in the synthesis strategy:

In synthesis A of Scheme 10.22, all of the isolated stereogenic centers and the first of the stereotriad were derived from the chiral pool. The stereotriad and the further center in 1,3-relationship were then elaborated by asymmetric induction from the first stereogenic center in a sequence of two aldol additions [44]. In synthesis B of Scheme 10.22, the left end of the stereotriad as well as the center next to the carbonyl group were taken from the chiral pool. The stereotriad was finished by an asymmetric aldol addition. The isolated stereogenic center with the methyl branch was derived from an asymmetric alkylation of an Evans enolate. The isolated stereogenic center in the lactone ring was generated in an asymmetric hetero-Diels-Alder cycloaddition. Both syntheses thus follow a building block oriented strategy combined with

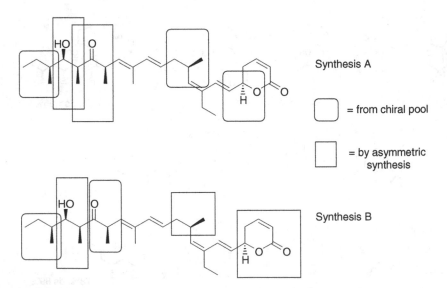

Scheme 10.22 Provenance of stereogenic centers in two syntheses of callystatin A

the recognition of patterns (the stereotriad) that dictated method-oriented sequences of steps.

In summary, when considering target structures that contain several neighboring stereogenic centers, try to assign these to subgroups, and look for congruence with known "patterns of stereogenic centers" for which established synthetic inroads exist, or which can be derived from the chiral pool. This kind of analysis usually leads to a synthesis plan that is partially building block oriented and partially method oriented.

Problems

10.1 Nonactin is one of the macrotetrolide antibiotics. It contains four entities of nonactic acid, pair-wise of opposite configuration. A synthesis of nonactin therefore requires access to each enantiomer of nonactic acid. Nonactic acid has four stereogenic centers, in an acyclic 1,2- and 1,3-distance and in a cyclic 1,4-disposition. Suggest a synthetic approach to nonactic acid (Scheme 10.23), that relies on recruiting the first stereogenic center from the chiral pool and deriving the remaining ones by asymmetric synthesis.

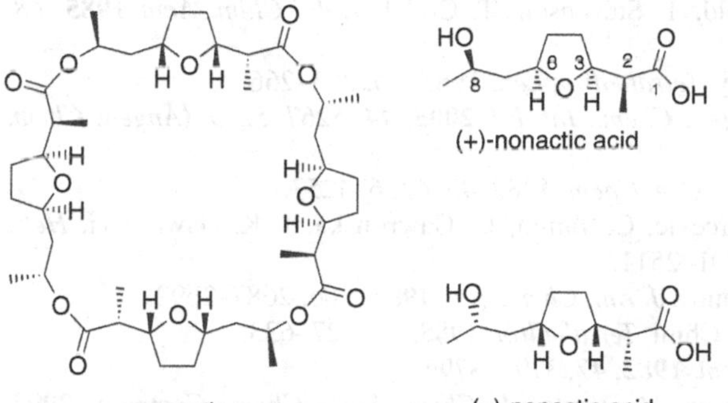

Scheme 10.23 Nonactin and nonactic acid

References

1. G. Quinkert, H. Stark, *Angew. Chem. Int. Ed. Engl.* **1983**, *22*, 637–655. (*Angew. Chem.* **1983**, *95*, 651–669.
2. A. Belan, J. Bolte, A. Fauve, J. G. Gourcy, H. Veschambre, *J. Org. Chem.* **1987**, *52*, 256–260.
3. E. J. Corey, R. K. Bakshi, S. Shibata, *J. Am. Chem. Soc.* **1987**, *109*, 5551–5553.
4. P. K. Jadhav, K. S. Bhat, P. T. Perumal, H. C. Brown, *J. Org. Chem.* **1986**, *51*, 432–439.

5. D. Hoppe, T. Hense, *Angew. Chem. Int. Ed. Engl.* **1997**, *36*, 2282–2316. (*Angew. Chem.* **1997**, *109*, 2376–2410).
6. J. W. Scott in *Asymmetric Synthesis* (Eds.: J. D. Morrison, J. W. Scott), Academic Press, New York, vol. 4, **1984**, pp. 1–226.
7. K. Mori, *Tetrahedron* **1975**, *31*, 3011–3012.
8. H. R. Schuler, K. N. Slessor, *Can. J. Chem.* **1977**, *55*, 3280–3287.
9. B. D. Johnston, K. N. Slessor, *Can. J. Chem.* **1979**, *57*, 233–235.
10. S. Takano, M. Yanase, M. Takahashi, K. Ogasawara, *Chem. Lett.* **1987**, *16*, 2017–2020.
11. K. Nakamura, M. Kinoshita, A. Ohno, *Tetrahedron* **1995**, *51*, 8799–8808.
12. A. Steinreiber, A. Stadler, S. F. Mayer, K. Faber, O. C. Kappe, *Tetrahedron Lett.* **2001**, *42*, 6283–6286.
13. B. H. Lipshutz, J. M. Servesko, *Angew. Chem., Int. Ed.* **2003**, *42*, 4789–4792. (*Angew. Chem.* **2003**, *115*, 4937–4940).
14. J. M. Brown in *Comprehensive Asymmetric Catalysis* (Eds.: E.-N. Jacobsen, A. Pfaltz, H. Yamamoto), Springer, Berlin, vol. 1, **1999**, pp. 121–182.
15. R. L. Halterman in *Comprehensive Asymmetric Catalysis* (Eds.: E.-N. Jacobsen, A. Pfaltz, H. Yamamoto), Springer, Berlin, vol. 1, **1999**, pp. 183–195.
16. W. Oppolzer, R. J. Mills, M. Réglier, *Tetrahedron Lett.* **1986**, *27*, 183–186.
17. W. Oppolzer, G. Poli, *Tetrahedron Lett.* **1986**, *27*, 4717–4720.
18. S. Rendler, M. Oestreich, *Angew. Chem., Int. Ed.* **2007**, *46*, 498–504. (*Angew. Chem.* **2007**, *119*, 504–510).
19. D. A. Evans, M. D. Ennis, D. J. Mathre, *J. Am. Chem. Soc.* **1982**, *104*, 1737–1739.
20. F. Lopez, A. J. Minnaard, B. L. Feringa, *Acc. Chem. Res.* **2007**, *40*, 179–188.
21. W. Oppolzer, P. Dudfield, T. Stevenson, T. Godel, *Helv. Chim. Acta* **1985**, *68*, 212–215.
22. F. E. Ziegler, A. Kneisley, *Tetrahedron Lett.* **1985**, *26*, 263–266.
23. C. Herber, B. Breit, *Angew. Chem., Int. Ed.* **2005**, *44*, 5267–5269. (*Angew. Chem.* **2005**, *117*, 5401–5403).
24. B. M. Trost, T. P. Klun, *J. Org. Chem.* **1980**, *45*, 4256–4257.
25. P. Mohr, N. Waespe-Sarcevic, C. Tamm, K. Gawronska, J. K. Gawronski, *Helv. Chim. Acta* **1983**, *66*, 2501–2511.
26. B. Cambou, A. M. Klibanov, *J. Am. Chem. Soc.* **1984**, *106*, 2687–2692.
27. R. Rossi, A. Carpita, M. Chini, *Tetrahedron* **1985**, *41*, 627–633.
28. P. E. Sonnet, *J. Org. Chem.* **1982**, *47*, 3793–3796.
29. A. G. Pepper, G. Procter, M. Voyle, *J. Chem. Soc., Chem. Commun.* **2002**, 1066–1067.
30. C. Arsene, S. Schulz, *Org. Lett.* **2002**, *4*, 2869–2871.
31. W. C. Still, J. C. Barrish, *J. Am. Chem. Soc.* **1983**, *105*, 2487–2489.
32. K. Suzuki, E. Katayama, G. Tsuchihashi, *Tetrahedron Lett.* **1984**, *25*, 2479–2482.
33. W. C. Still, J. A. Schneider, *Tetrahedron Lett.* **1980**, *21*, 1035–1038.
34. Y. Yamamoto, N. Asao, *Chem. Rev.* **1993**, *93*, 2207–2293.
35. C. J. Cowden, I. Paterson, *Org. React.* **1997**, *51*, 1–200.
36. R. W. Hoffmann, *Angew. Chem. Int. Ed. Engl.* **1987**, *26*, 489–503. (*Angew. Chem.* **1987**, *99*, 503–517).
37. I. Paterson, J. A. Channon, *Tetrahedron Lett.* **1992**, *33*, 797–800.
38. I. Paterson, R. D. Tillyer, *Tetrahedron Lett.* **1992**, *33*, 4233–4236.
39. J. A. Marshall, G. M. Schaaf, *J. Org. Chem.* **2001**, *66*, 7825–7831.
40. O. Arjona, R. Menchaca, J. Plumet, *J. Org. Chem.* **2001**, *66*, 2400–2413.

41. R. Stürmer, R. W. Hoffmann, *Chem. Ber.* **1994**, *127*, 2519–2526.
42. R. W. Hoffmann, R. Stürmer, *Chem. Ber.* **1994**, *127*, 2511–2518.
43. R. W. Hoffmann, R. Stürmer *Towards Erythronolides, Efficient Synthesis of Contiguous Stereocenters* in *Antibiotics and Antiviral Compounds, Chemical Synthesis and Modification* (Eds.: K. Krohn, H. Kirst, H. Maas), VCH Verlagsges., Weinheim, **1993**, pp. 103–110.
44. M. Kalesse, M. Christmann, *Synthesis* **2002**, 981–1003.

Chapter 11
Enjoying the Art of Synthesis

Abstract Weaknesses and strong points of a synthesis (plan) become apparent by comparing several syntheses of the same target. Such a comparison highlights surprising solutions—the art of synthesis—in contrast to standard approaches.

The planning of syntheses is in the middle of a transformation from an art into a technique. When confronted with targets of unusual structure, surprising solutions are still called for. This means that at any time creative and artistic synthesis schemes have to be put forward, when the systematic (scholastic) approach fails or turns out to be too lengthy.

Art contributes to and is part of culture, which you may (and should) enjoy. The same can be said of exemplary syntheses. After having looked at various systematic ways to deal with the problems of planning a synthesis, pointing out the art in outstanding syntheses is the best conclusion. In order to truly enjoy art, one must be knowledgeable about art. A first glance at a painting of Picasso or of Klee will arouse one's interest in the kind of presentation at a superficial level. Only continuing exposure to modern art will let one recognize certain recurring details and patterns in paintings of Picasso or Klee, because one's eyes have been trained by the preceding studies. To see more and to recognize more is what makes looking at pictures enjoyable. The same holds for those who have studied the underlying concepts of synthesis planning and now look at published syntheses. The reader who has made his or her way through this treatise should be knowledgeable enough to recognize many familiar aspects that were presented in the preceding chapters. With the insight gained in this manner, one should be able to discern creative syntheses from boring ones, i.e., to enjoy the real art of synthesis.

To this end, there follows a compilation of syntheses, each of which has some remarkable feature. The reader is encouraged to rediscover familiar elements and to examine in depth the way the synthesis was conceived. As the guide in an art museum helps the visitor to appreciate the features of the

R.W. Hoffmann, *Elements of Synthesis Planning*,
DOI 10.1007/978-3-540-79220-8_11, © Springer-Verlag Berlin Heidelberg 2009

masterpieces on display, our comments should help the reader's appreciation of the following examples of outstanding synthesis.

11.1 Strychnine

Strychnine was declared by Robinson to be the most complex compound relative to its size [1]. The elucidation of the structure of strychnine, in a period in which neither NMR-spectroscopy nor X-ray crystal structure analysis were available, constitutes one of the eminent cultural achievements of chemistry. The challenge of its synthesis could not have been accepted by a person of lesser standing than R. B. Woodward, who in the middle of the last century was beyond dispute the genius of synthesis. Woodward succeeded in the first synthesis of strychnine (Scheme 11.1) [2]. This benchmark synthesis appears by today's standards as a careful and somewhat reluctant approach to a solution of the problem. Yet the ingenuity of Woodward becomes manifest in two features: construction of the quaternary stereogenic center as the pivotal point of three rings, and the concept of introducing a veratryl substituent on the indole ring and its unprecedented use.

The veratryl moiety was conceived by Woodward to serve as a latent functionality for a branching point with one unsaturated C_2- and one unsaturated C_3-chain. This was revealed in step 6 by a daring chemoselective (presence of the indole moiety!) and position-specific ozonolytic cleavage of the aromatic ring between the carbons bearing the methoxy substituents generating two ester functionalities. The compact nature of the veratryl group allowed the start of the synthesis from an easily accessible starting material.

Steps 1 through 4 of the synthesis address the generation of the pivotal quaternary stereogenic center at the fusion point of three rings. This goal was attained by starting with a classical tryptamine synthesis (steps 1 and 2) followed by condensation with ethyl glyoxalate (step 3). The pyrrolidine ring was closed and the quaternary stereogenic center was established in a Mannich-cyclization initiated by tosyl chloride (step 4). Refunctionalization (step 5) secured the conquered structure. As indicated above, step 6 unravelled the veratryl ring, with the result that one side chain immediately cyclized to form the next (pyridone) ring, shifting the carbon-carbon double bonds to their most stable positions. This is the key step of the total synthesis, marking a considerable increase in complexity and setting the functionality for forming the next ring in step 8. First, though, the tosyl group from the pyrrolidine nitrogen had to be removed in step 7. Step 8 marks the formation of the next ring by a Dieckmann cyclization. This entailed an enolized carbonyl group in the newly-formed ring as a surplus functionality.

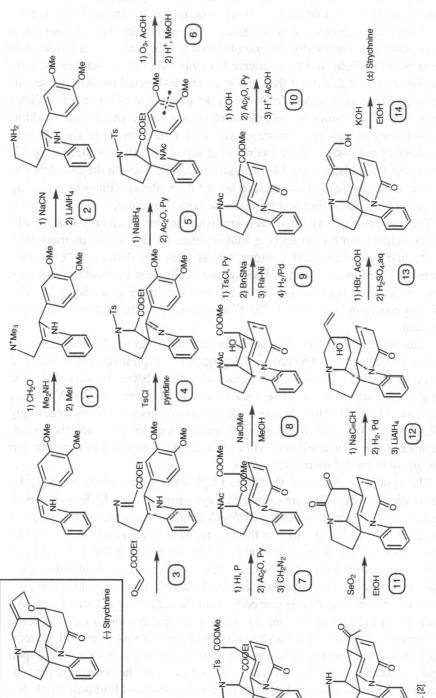

Ref. [2]

Scheme 11.1 Strychnine synthesis by R. B. Woodward

Its reductive removal necessitated a total of four operations (step 9)! The next transformation, that of a methoxycarbonyl group into a methyl ketone (step 10) via intramolecular acetyl transfer in an anhydride is surprising and unconventional. The methyl ketone obtained in this manner was oxidized in the α-position. Followed by epimerization this led to the closure of the next ring (step 11). At this point the carbon atoms designated to form the seven-membered ring were added, followed by an adjustment of the functionalities (reduction of the amide to an amine, reduction of the pyridone to a dihy-dropyridone) in step 12. The stage was set for the closure of the last ring by effecting a 1,3-shift of the hydroxyl group in the allylic system, step 13. Treatment with a base (step 14) brought the double bond in the dihydropyri-done into conjugation, allowing it to accept the hydroxyl function in closing the seven-membered ring and to reach the target structure.

The synthesis of strychnine accomplished in this manner was a breath-taking achievement. Considering that it relies solely on the methodology of the first half of the 20th century, this is what one defines as "classical" chemistry. One notes that the Dieckmann cyclization necessitated four re-functionalization operations, something one would try to avoid today. But this appears permissible for a "first" synthesis of such a formidable target molecule.

Almost thirty years had to pass before chemists gained the confidence to start a new attack on strychnine. They hoped to capitalize on the widened scope of synthetic methods to realize a perhaps better solution [3]. The most remarkable one of these endeavours is the synthesis by V. Rawal (Scheme 11.2) [4]. This synthesis is impressive because of its seemingly effortless late increase in complexity, and the manner in which the seven-membered ring was anellated. This set a standard which was incorporated into many of the following syntheses.

The synthesis of Rawal (Scheme 11.2) starts rather conventionally by elaborating the pyrroline ring through a cyclopropane-aldehyde and an aza-analogue of a vinyl-cyclopropane/cyclopentene rearrangement (steps 1 and 2). The resulting enamine is then protected as an enamide in step 3. The amino function on the aryl residue is in turn generated and converted to an imine by reaction with a strategically designed branched, unsaturated alde-hyde. Once the imine is acylated (step 4), the double bonds are shifted to gen-erate a diene that is perfectly poised to form the central ring of strychnine in a Diels-Alder cycloaddition (step 5). Steps 4 and 5 thus mark a rapid increase in complexity based on a bicyclization strategy. At this point, all protect-ing groups were convergently cleaved with trimethylsilyl iodide, whereupon acidic workup formed the next ring, the lactam. Now the remaining pyrroli-dine nitrogen was alkylated with a specifically designed building block that contained the functionality to form the next ring in addition to the skeletal

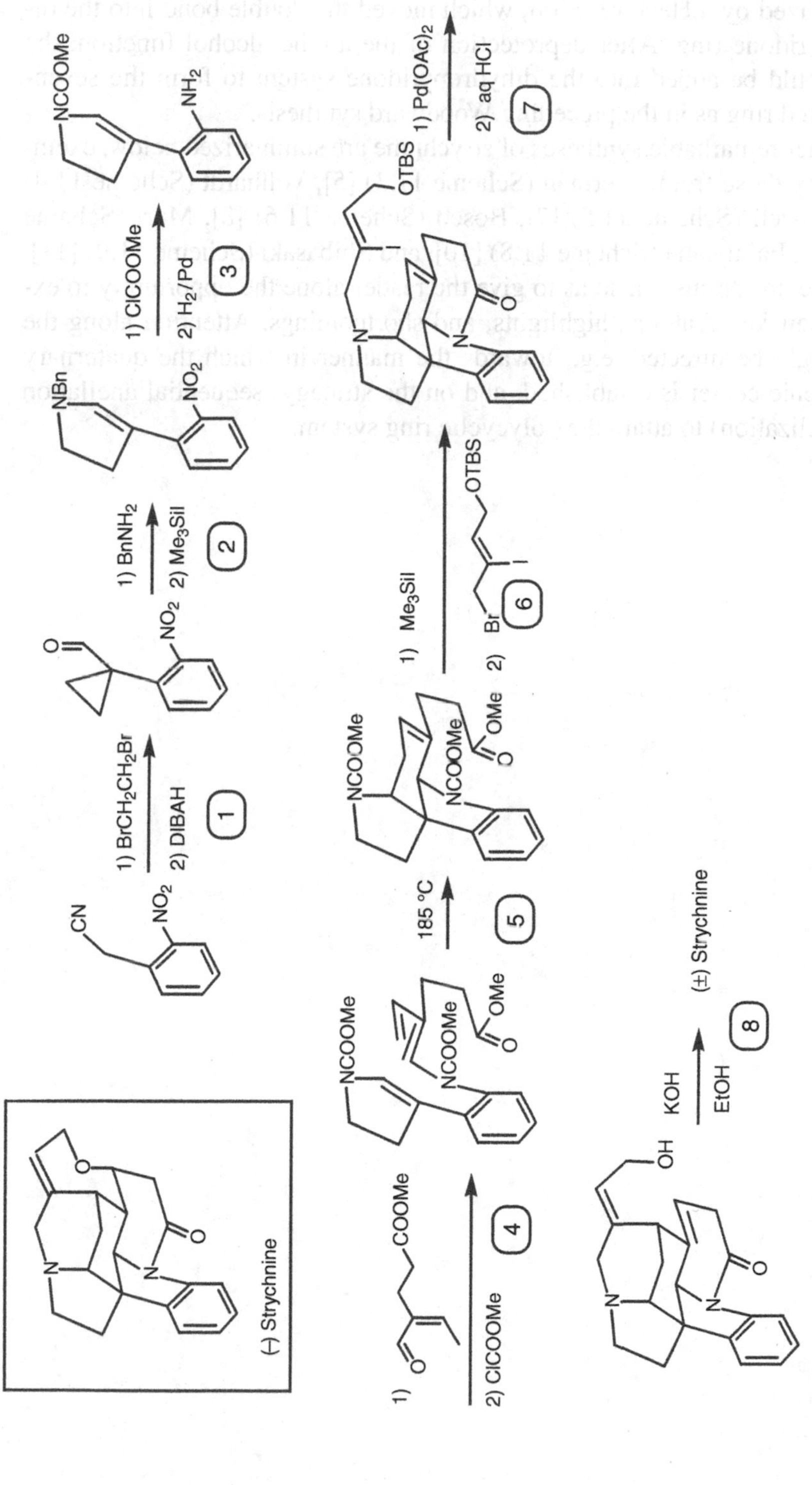

Scheme 11.2 Strychnine synthesis by V. H. Rawal

Ref. [4]

atoms to form the seven-membered ring (step 6). Closure of the next ring was realized by a Heck reaction, which moved the double bond into the dihydropyridone ring. After deprotection of the allylic alcohol function, the latter could be added into the dihydropyridone system to form the seven-membered ring as in the preceding Woodward synthesis.

Further remarkable syntheses of strychnine are summarized below, exemplified by those from Overman (Scheme 11.3) [5], Vollhardt (Scheme 11.4) [6], Bodwell (Scheme 11.5) [7], Bosch (Scheme 11.6) [8], Mori (Scheme 11.7) [9], Fukuyama (Scheme 11.8) [10], and Shibasaki (Scheme 11.9) [11]. They are not discussed, so as to give the reader alone the opportunity to explore their key features, highlights, and shortcomings. Attention along the way might be directed, e.g., towards the manner in which the quaternary stereogenic center is established, and on the strategy (sequential anellation or bicyclization) to attain the polycyclic ring system.

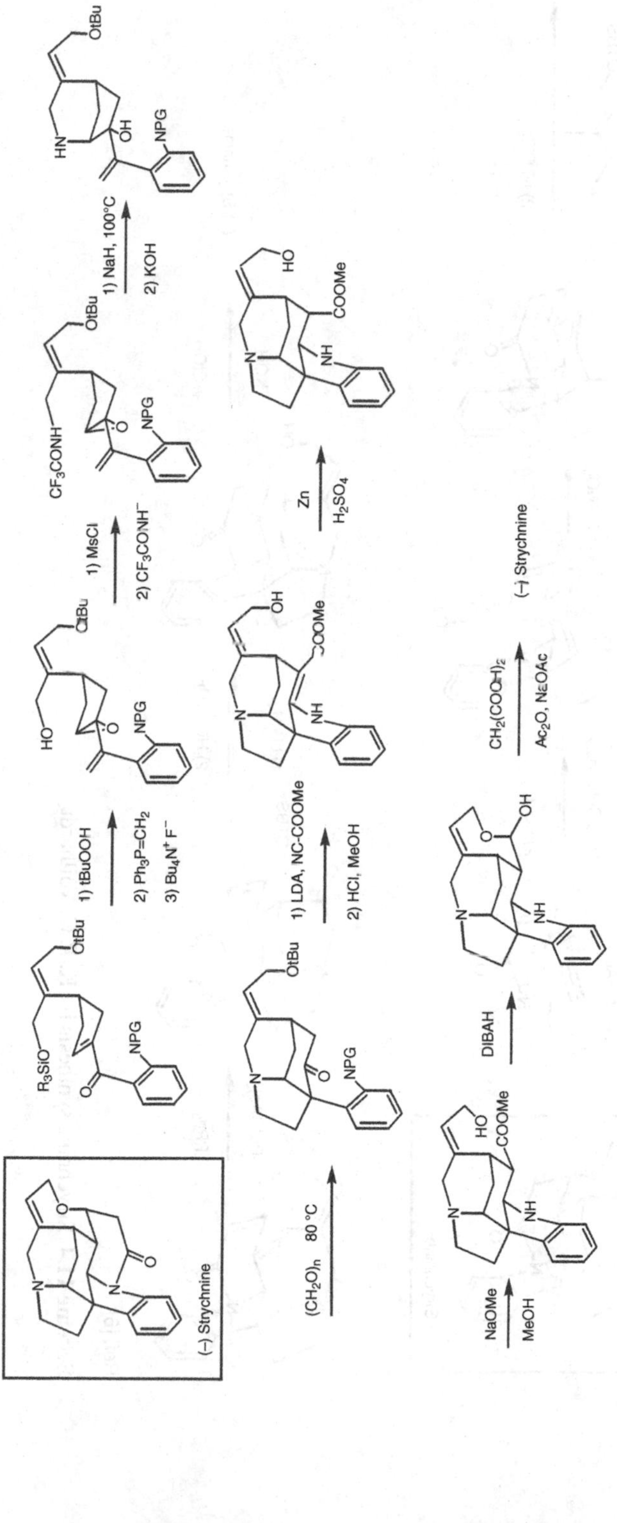

Scheme 11.3 Strychnine synthesis by L. E. Overman

Ref. [5]

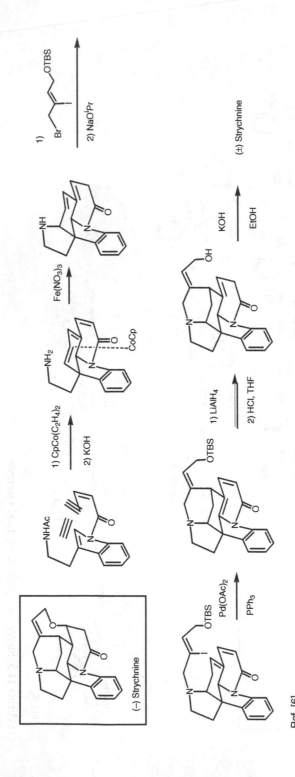

Ref. [6]

Scheme 11.4 Strychnine synthesis by K. P. C. Vollhardt

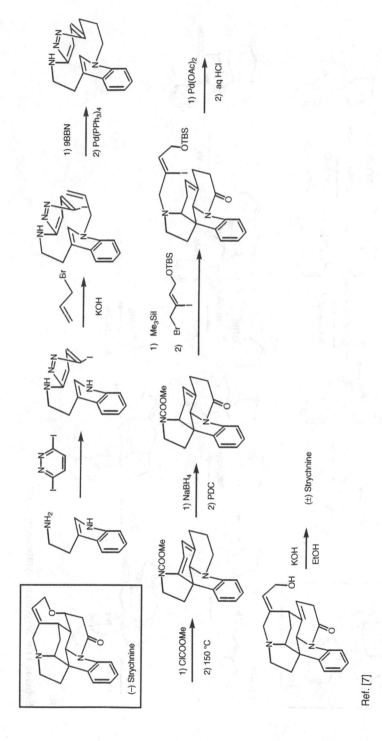

Ref. [7]

Scheme 11.5 Strychnine synthesis by G. J. Bodwell

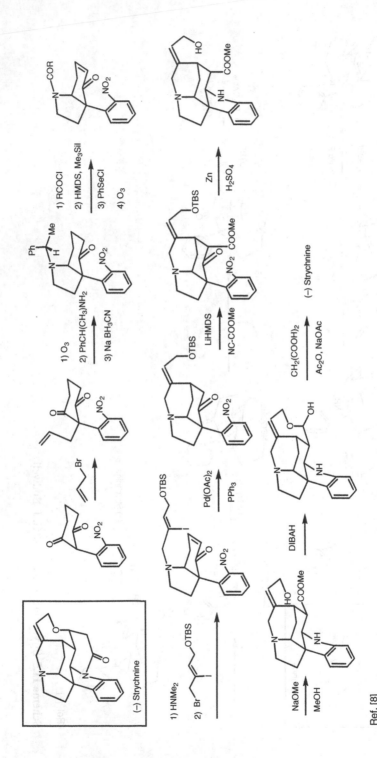

Scheme 11.6 Strychnine synthesis by J. Bosch

Ref. [8]

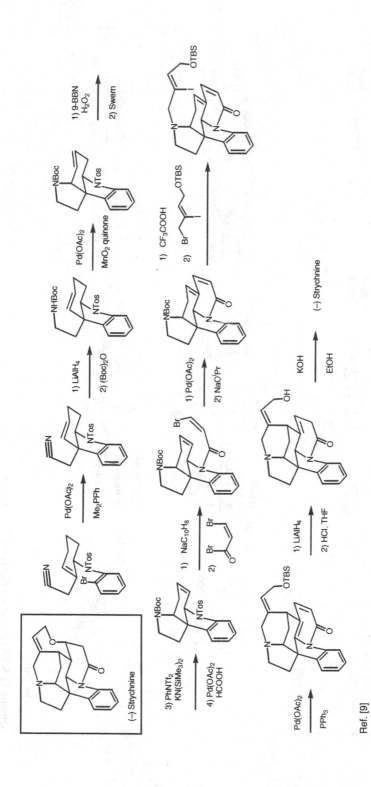

Scheme 11.7 Strychnine synthesis by M. Mori

Ref. [9]

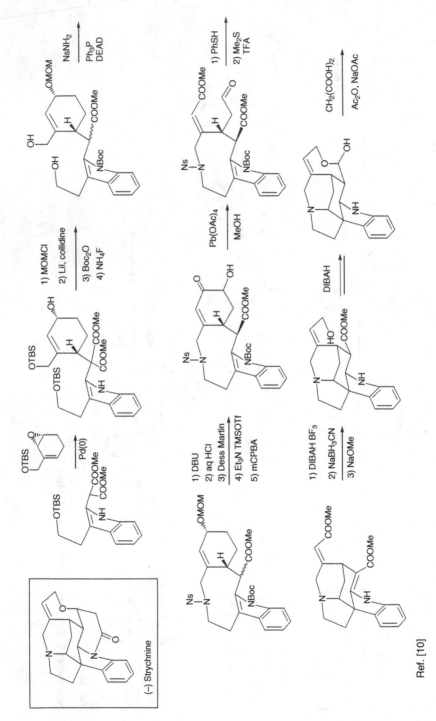

Ref. [10]

Scheme 11.8 Strychnine synthesis by T. Fukuyama

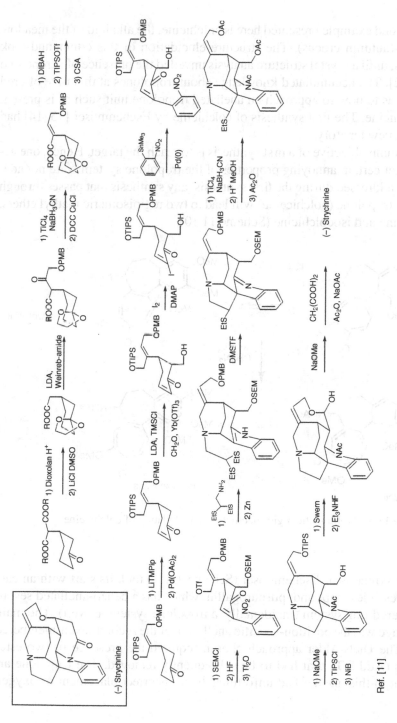

Scheme 11.9 Strychnine synthesis by M. Shibasaki

Ref. [11]

11.2 Colchicine

The second example presented here is colchicine, the alkaloid of the meadow saffron (autumn crocus). The structure elucidation of this compound took decades, until a crystal structure analysis unveiled the presence of a tropolone ring [12]. The accumulated knowledge about tropolones at that time offered no hint as to how to approach an anellated tropolone unit such as is present in colchicine. The first synthesis of colchicine by Eschenmoser [13, 14] had to chart new territory.

The main objective of a first synthesis is to reach the target. Hence, one accepts that certain annoying properties of the tropolone system were not adequately addressed during the first synthesis: any synthesis that passes through the free tropolone (colchiceine) will lead to two regioisomeric methyl ethers, colchicine and isocolchicine (Scheme 11.10).

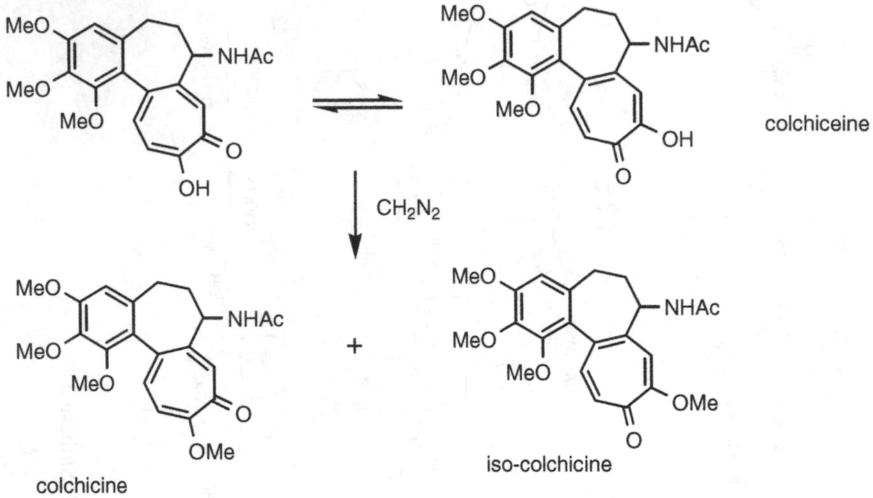

Scheme 11.10 Problematic regioselectivity on methylation of colchiceine

The synthesis of Eschenmoser (Scheme 11.11) took its start with an easily accessible compound, purpurogallin, which has a benzo-anellated seven-membered ring system (incidentally, a tropolone system as well). The main challenge was the position-specific anellation of the second seven-membered ring. The Diels-Alder approach chosen required the presence of two ester groups (Add FG), that had to be subsequently removed. The tropolone arrived at in this manner had unfortunately an incorrect placement of oxygen functionalities! The necessary correction entailed many steps and substantial losses of material.

This first synthesis of colchicine by Eschenmoser was followed by 14 further syntheses, which have been comprehensively discussed recently [15]. Four of these syntheses, those by Woodward (Scheme 11.12) [16], Evans (Scheme 11.13) [17], Schmalz (Scheme 11.14) [15, 18, 19], and by Cha (Scheme 11.15) [20] will be presented below. The syntheses by Woodward, Evans, and Schmalz likewise do not offer a solution to the regioselectivity problem associated with colchiceine. That was finally attained by the synthesis of Cha, which is compromised, however, by a protecting group dance from Boc to acetyl and back to Boc at the amino group.

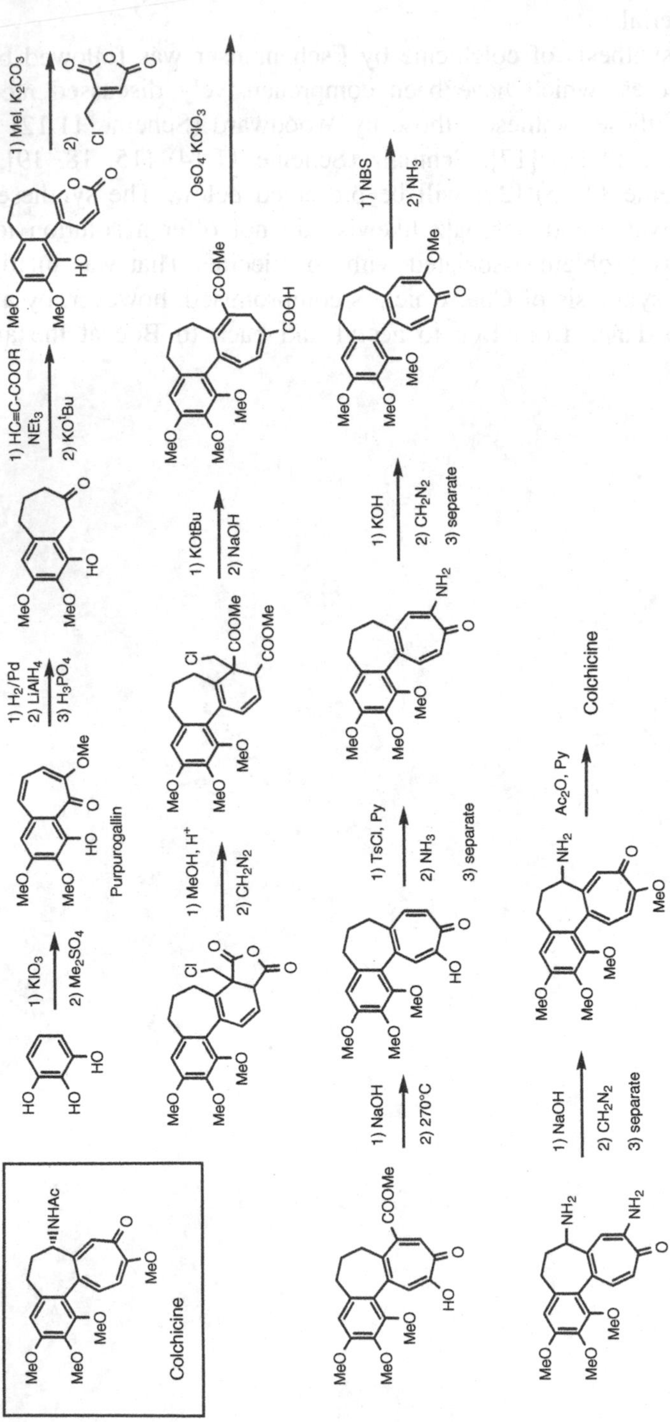

Scheme 11.11 Colchicine synthesis by A. Eschenmoser

Refs. [13, 14]

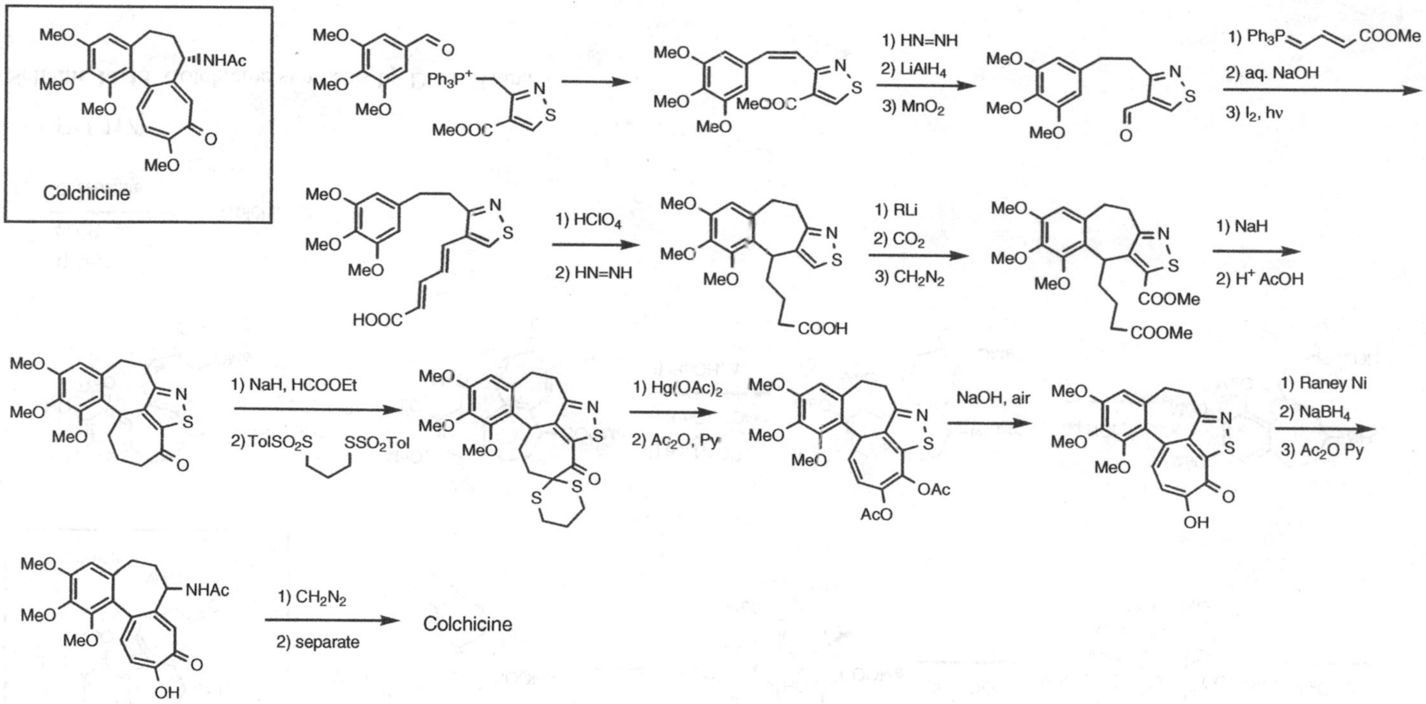

Scheme 11.12 Colchicine synthesis by R. B. Woodward

Ref. [16]

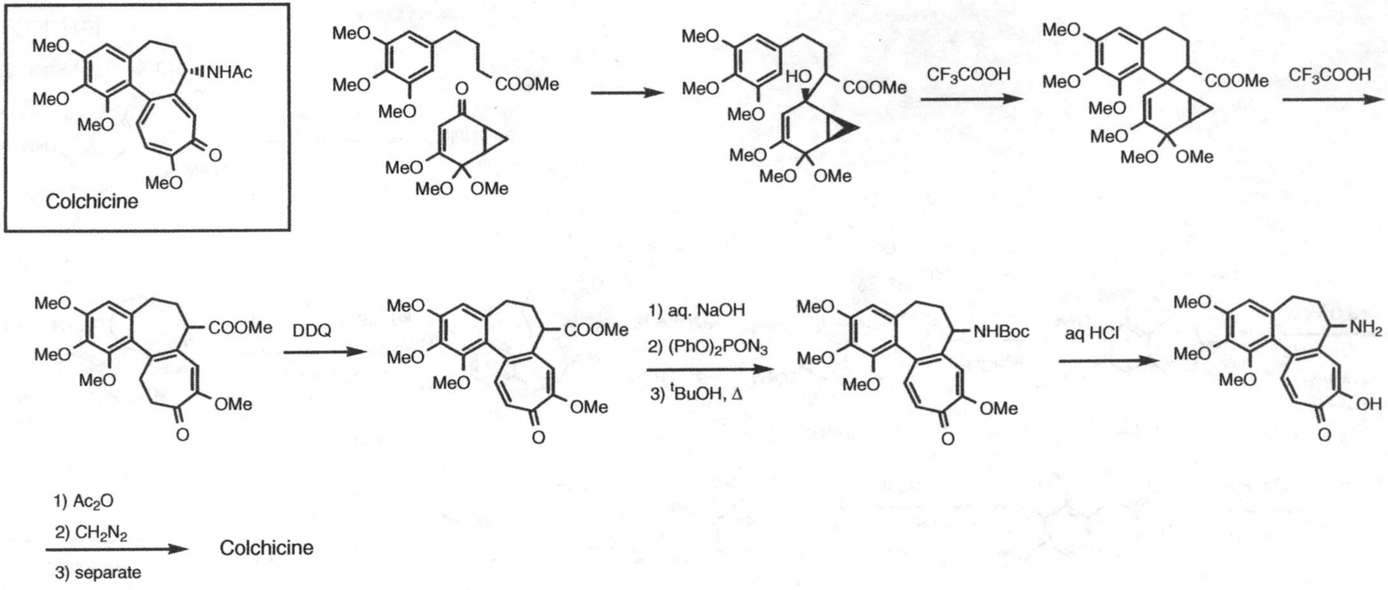

Scheme 11.13 Colchicine synthesis by D. A. Evans

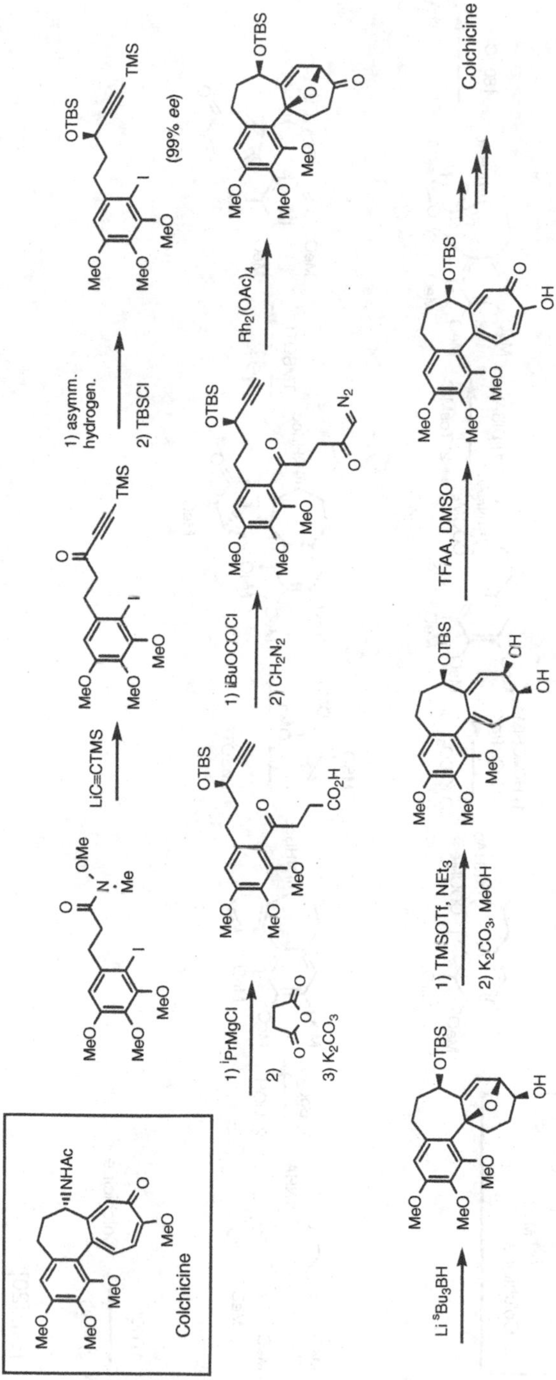

Scheme 11.14 Colchicine synthesis by H. G. Schmalz

Refs. [18, 19]

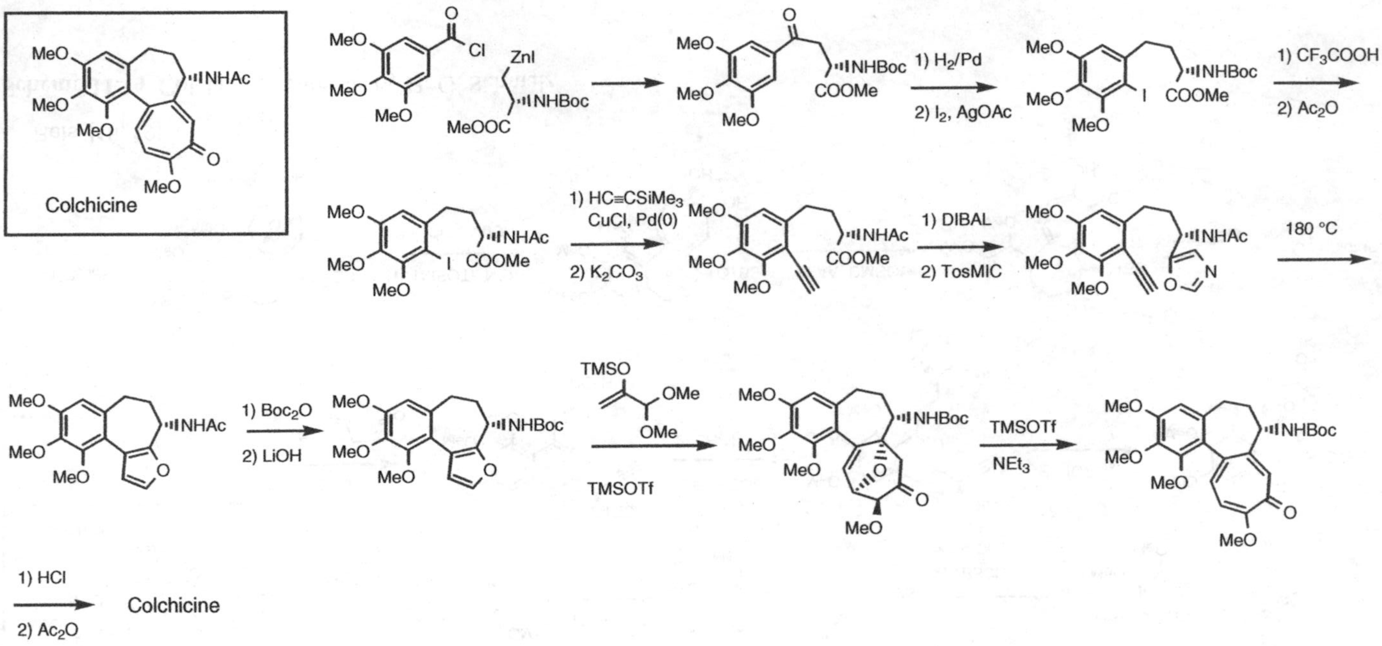

Ref. [20]

Scheme 11.15 Colchicine synthesis by J. K. Cha

11.3 Dysidiolide

Dysidiolide is a terpene of current interest in medicinal chemistry. Its structure invites creative approaches. The first synthesis by the Corey group [21]
(Scheme 11.16) starts from a readily available chiral building block with one
stereogenic center. The striking fact is that this stereogenic center appears
nowhere in the target structure. It serves as an auxiliary to derive the other
stereogenic centers and ultimately it had to be removed in an impressive
way.

The subsequent syntheses by Boukouvalas (Scheme 11.17) [22] and by
Danishefsky (Scheme 11.18) [23] focus on the double bond in the
six-membered ring, which allows a Diels-Alder approach. That however, requires activating groups at the dienophile that have to be removed in additional steps. Scheme 11.19 presents a combination of the syntheses by
Forsyth [24] and by Maier [25]. They, too, implement a Diels-Alder reaction,
again requiring an activating group on the dienophile. That leads in the end
to a side chain that is short by one carbon atom, a fact that has to be corrected
by a sequence of further steps.

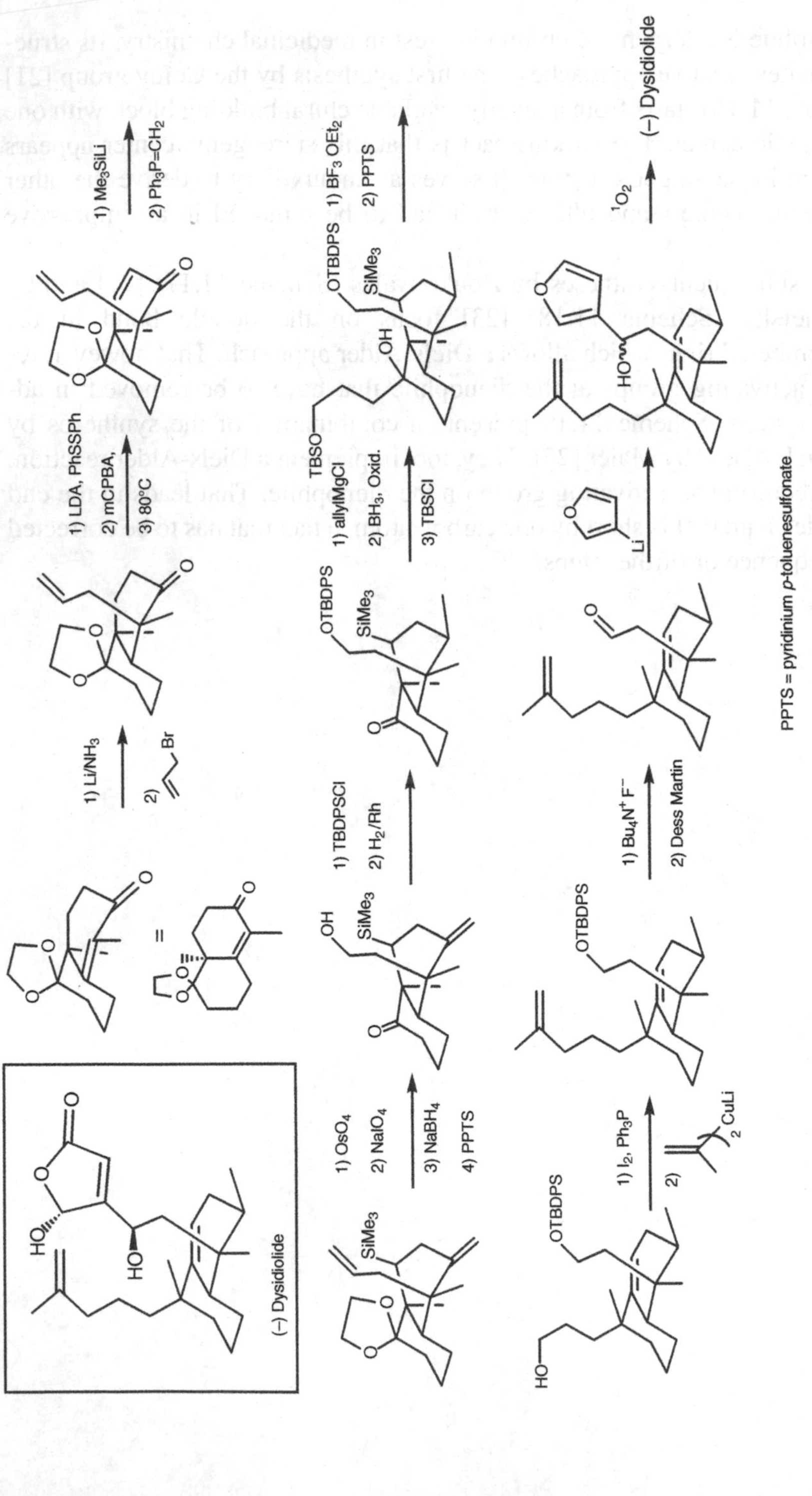

PPTS = pyridinium *p*-toluenesulfonate

Ref. [21]

Scheme 11.16 Dysidiolide synthesis by E. J. Corey

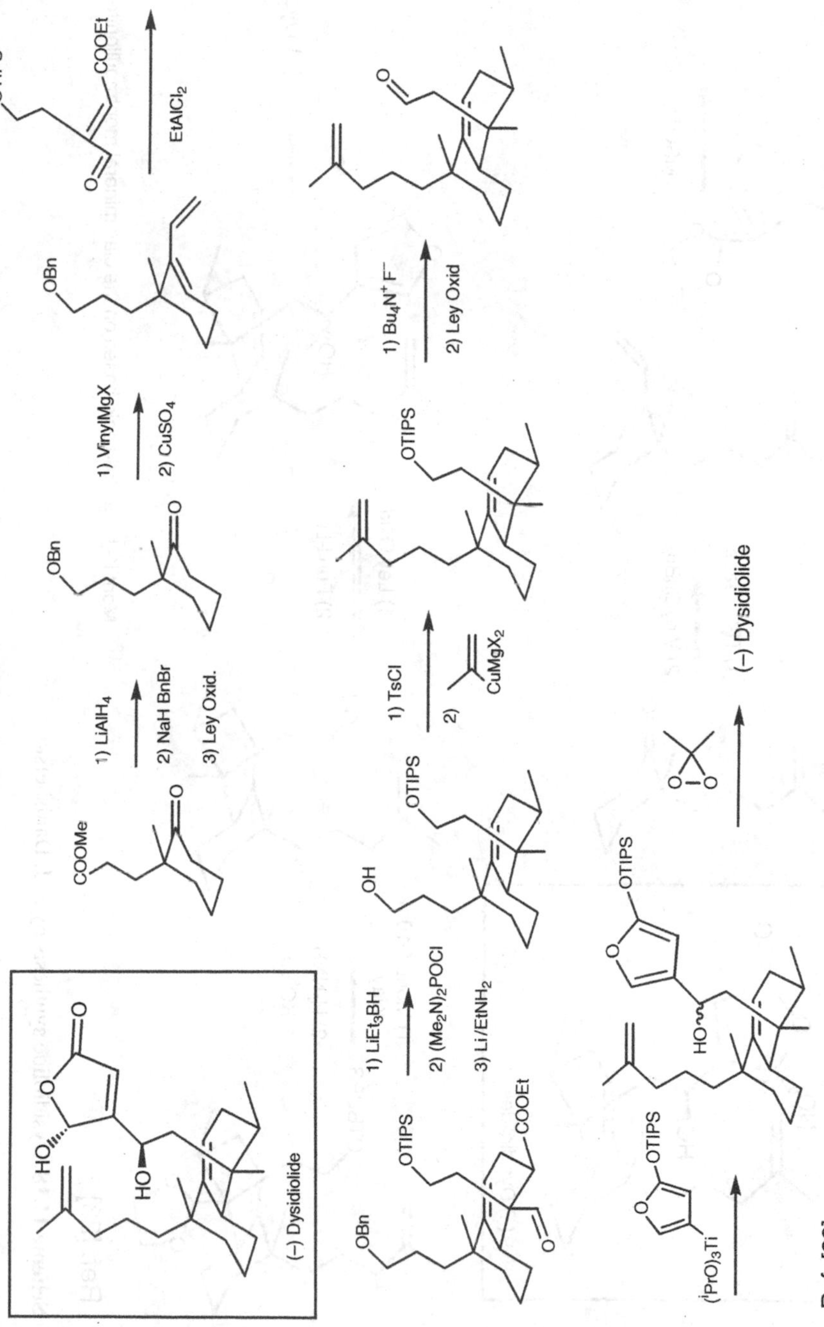

Scheme 11.17 Dysidiolide synthesis by J. Boukouvalas

Ref. [22]

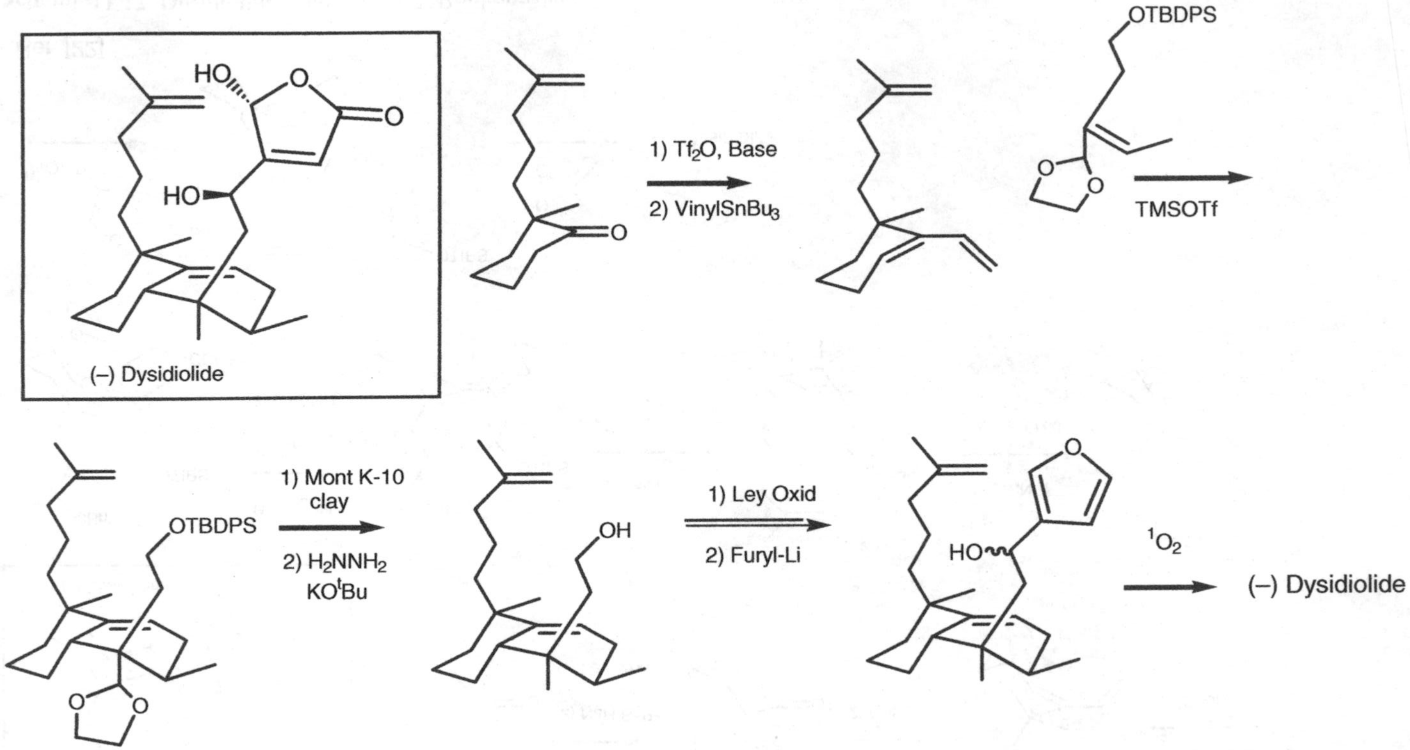

Mont K-10 = Solid acid based on the clay mineral montmorillonite

Ref. [23]

Scheme 11.18 Dysidiolide synthesis by S. J. Danishefsky

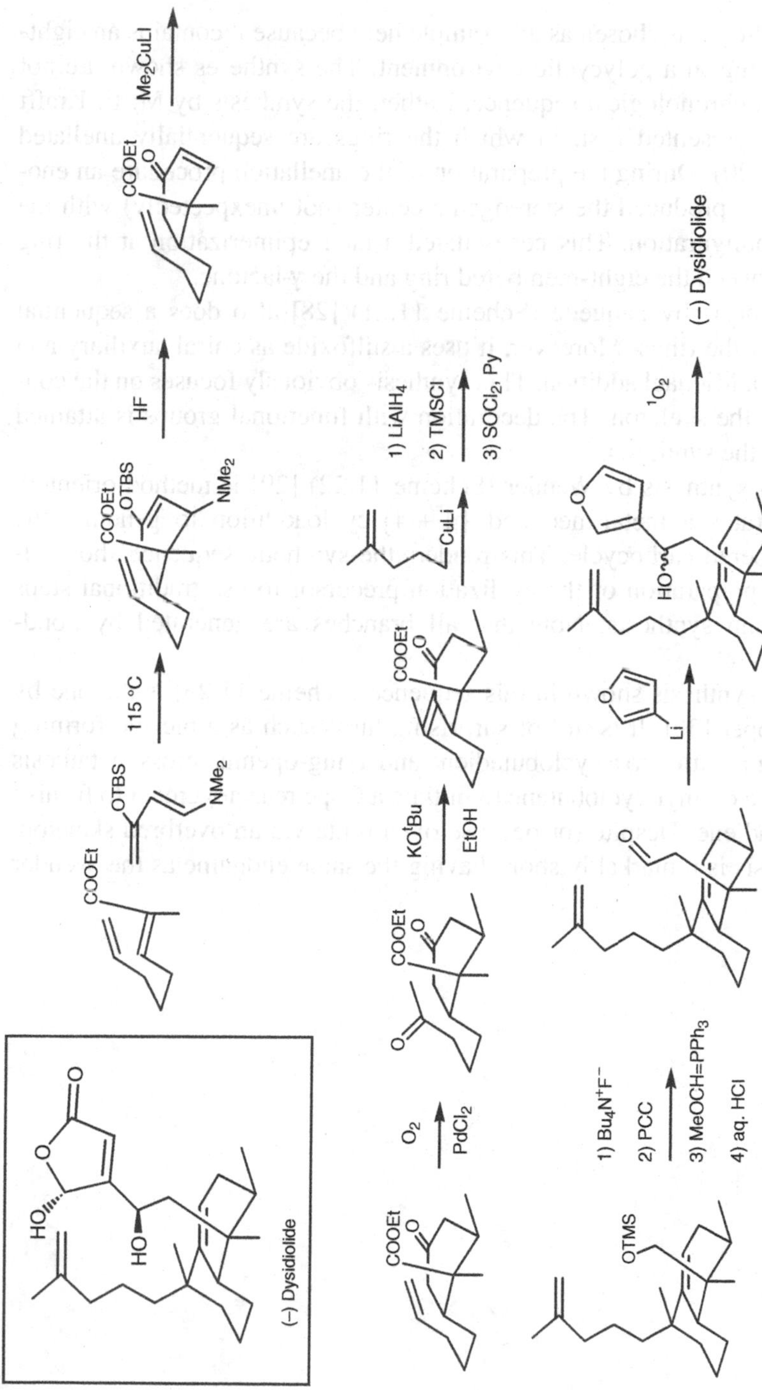

Refs. [24, 25]

Scheme 11.19 Dysidiolide synthesis by Forsyth and Maier

11.4 Asteriscanolide

Asteriscanolide was chosen as an example here because it contains an eight-membered ring in a polycyclic environment. The syntheses shown are not ordered in a chronological sequence. Rather, the synthesis by M. E. Krafft [26, 27], is presented first, in which the rings are sequentially anellated (Scheme 11.20). During the preparation of the anellation procedure an enolate allylation produced the stereogenic center (not unexpectedly) with the incorrect configuration. This necessitated a later epimerization at the ring juncture between the eight-membered ring and the γ-lactone.

The synthesis by Paquette (Scheme 11.21) [28] also does a sequential anellation of the rings. Moreover, it uses a sulfoxide as chiral auxiliary and to facilitate a Michael addition. This synthesis obviously focuses on the construction of the skeleton. The decoration with functional groups is attained only late in the synthesis.

The next synthesis by Wender (Scheme 11.22) [29] is method-oriented. It demonstrates a metal-mediated {4 + 4}-cycloaddition to generate the eight-membered carbocycle. This renders the synthetic sequence short, allowing the preparation of the cyclization precursor to use traditional steps of open-chain synthesis. Note that all branches are generated by bond-formation.

The last synthesis shown in this sequence (Scheme 11.23) is the one by M. L. Snapper [30]. It is full of surprising turns such as a bicycle-forming Diels-Alder addition to a cyclobutadiene and a ring-opening cross metathesis to generate a divinyl-cyclobutane to initiate a Cope rearrangement to furnish a cyclooctadiene. Despite (or because of) a route via an overbred skeleton, this synthesis is remarkably short, having the same endgame as the Wender synthesis.

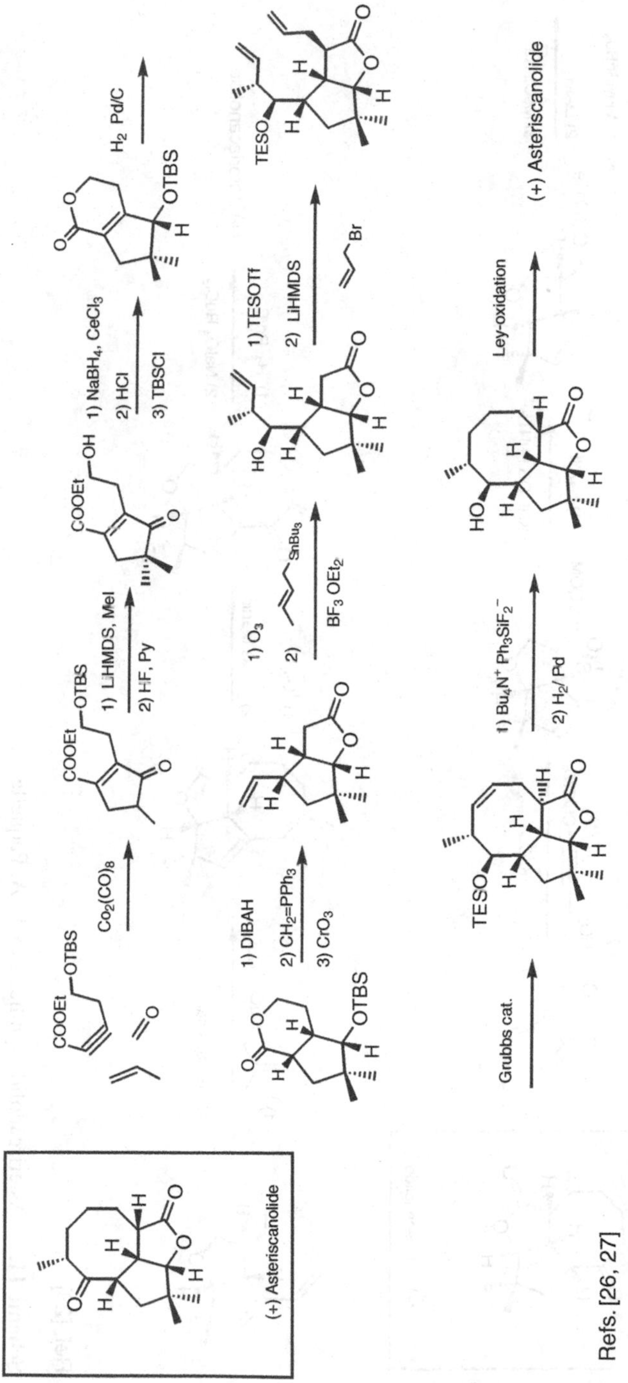

Refs. [26, 27]

Scheme 11.20 Asteriscanolide synthesis by M. E. Krafft

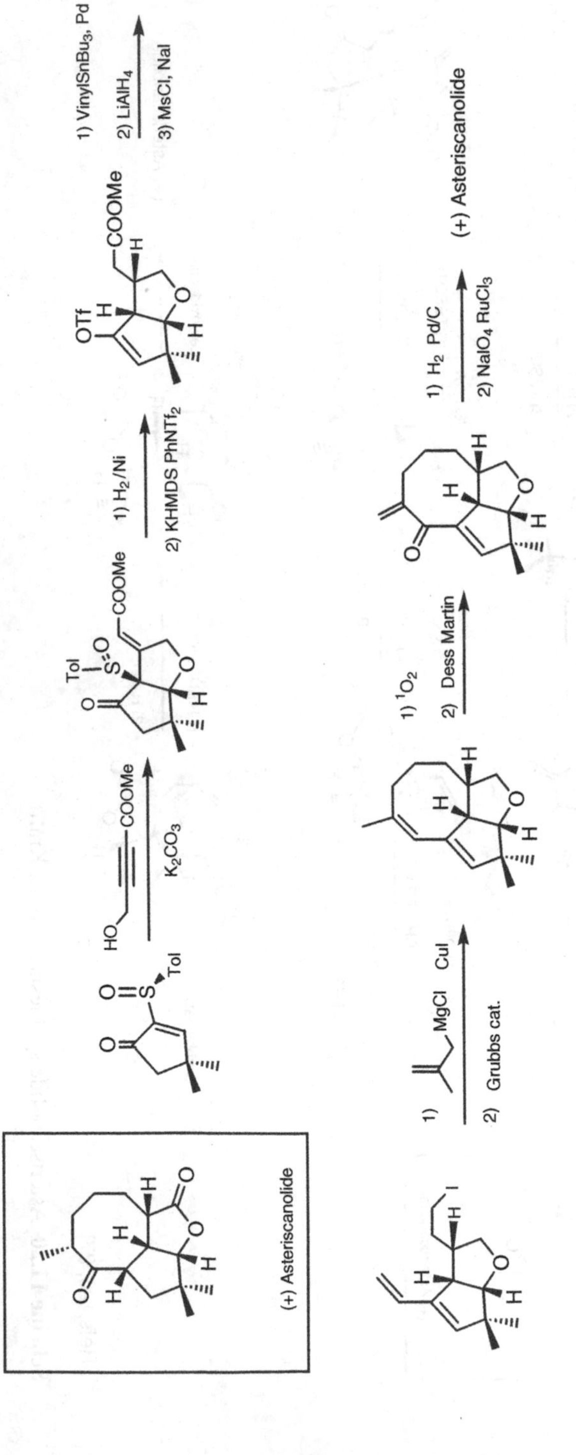

Ref. [28]

Scheme 11.21 Asteriscanolide synthesis by L. A. Paquette

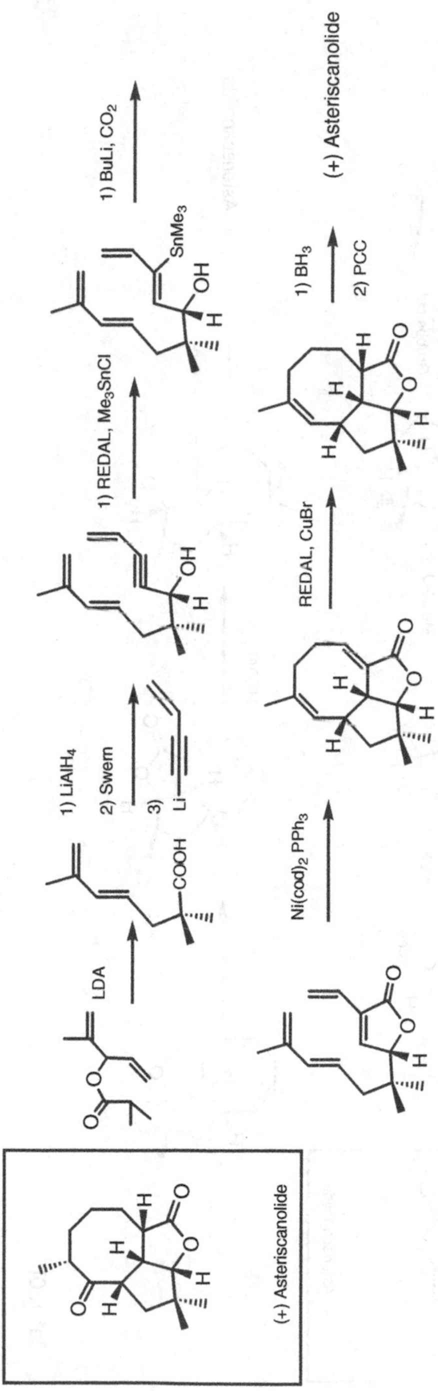

Ref. [29]

Scheme 11.22 Asteriscanolide synthesis by P. A. Wender

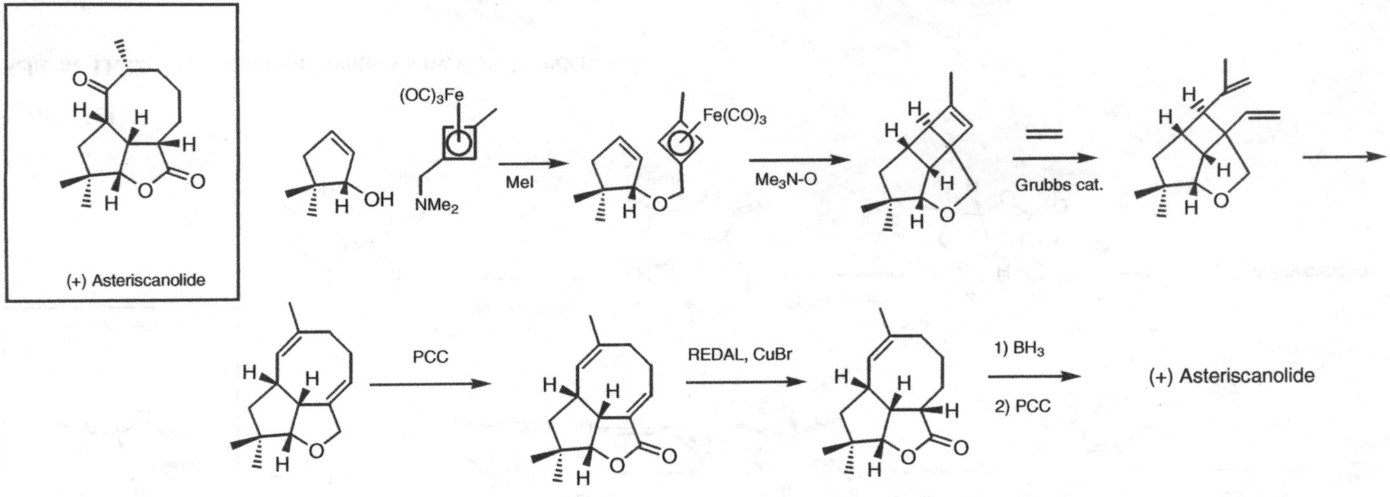

Ref. [30]

Scheme 11.23 Asteriscanolide synthesis by M. L. Snapper

11.5 Lepadiformine

Lepadiformine is a not-too-complicated tricyclic alkaloid, which finds some attention in the synthetic community [31]. The first synthesis by Funk [32] constructs the rings sequentially (Scheme 11.24). The functionality initially needed to facilitate the Diels-Alder addition has been cleverly used to build the pyrrolidine ring.

The synthesis by Weinreb (Scheme 11.25) [33, 34] is method oriented with respect to oxidation of the pyrrolidine ring by a radical translocation step. The synthesis tried to feature a latent hydroxyl group in the form of an allyldimethylsilyl moiety. Unfortunately, the latter group appeared not to be compatible with a number of subsequent steps. Because of this the hydroxyl group had to be unveiled early on.

The next synthesis by Kibayashi (Scheme 11.26) [35, 36, 37] shows a near miss of a bicyclization strategy, as the acylimmonium intermediate was generated under solvolytic conditions (using formic acid as the solvent). Thus, a formate was generated in a nonstereospecific manner that had to be saponified, requiring the additional adjustment of stereochemistry. Reflecting on this route, it appeared attractive to succeed in a bicyclization reaction. This appears to be possible in altering the acylimmonium ion to a simple immonium ion, which should be amenable to a Diels-Alder addition [38]. This led to the proposal shown in Scheme 11.27; for a key step see reference [39].

Lepadiformine nevertheless continues to be a target of synthetic endeavours [41, 42, 43, 44]. Perhaps the study of the syntheses presented here will encourage the reader to come up with shorter or more efficient proposals for this and other compounds. Their presentation was intended to give the reader the chance of enjoying the art of synthesis.

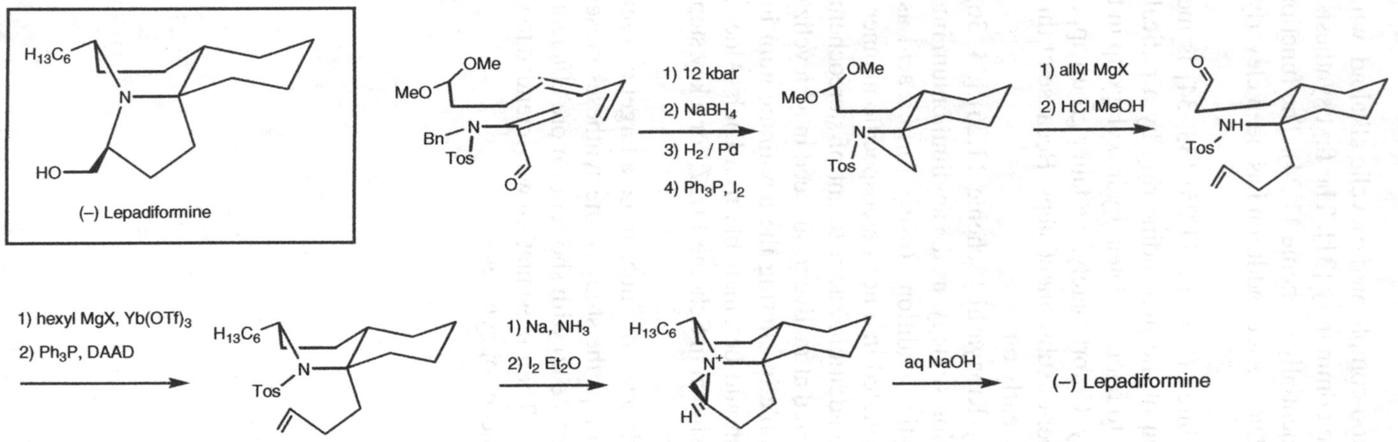

Ref. [32]

Scheme 11.24 Lepadiformine synthesis by R. L. Funk

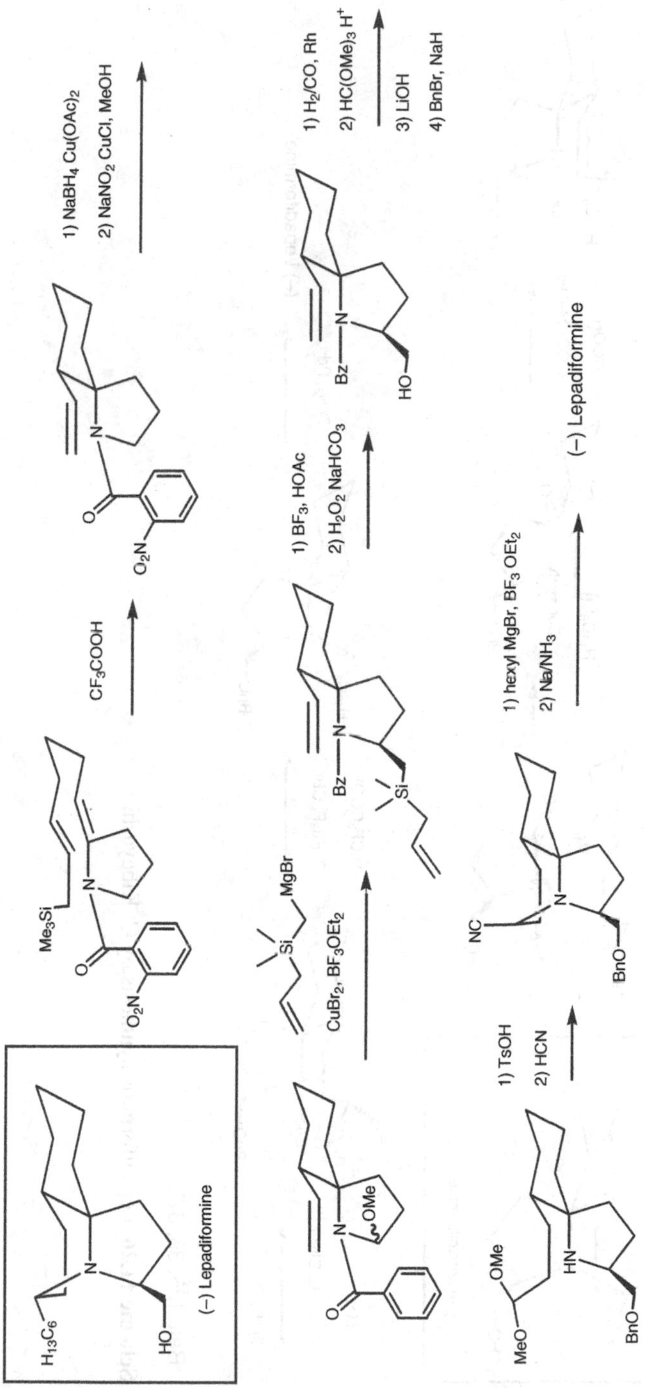

Refs. [33, 34]

Scheme 11.25 Lepadiformine synthesis by S. M. Weinreb

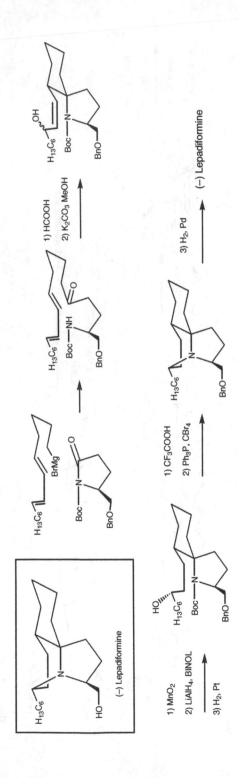

Refs.[35, 36, 37]

Scheme 11.26 Lepadiformine synthesis by C. Kibayashi

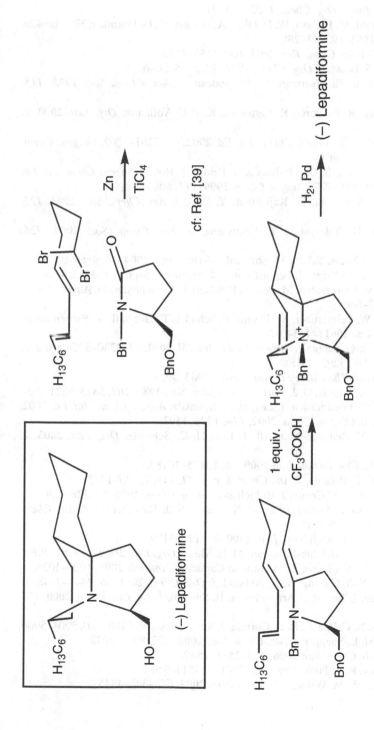

Scheme 11.27 Lepadiformine synthesis proposed by R. W. Hoffmann-group

References

1. R. Robinson, *Progr. Org. Chem.* **1952**, *1*, 1–21.
2. R. B. Woodward, M. P. Cava, W. D. Ollis, A. Hunger, H. U. Daeniker, K. Schenker, *Tetrahedron* **1963**, *19*, 247–288.
3. J. Bonjoch, D. Sole, *Chem. Rev.* **2000**, *100*, 3455–3482.
4. V. H. Rawal, S. Iwasa, *J. Org. Chem.* **1994**, *59*, 2685–2686.
5. S. D. Knight, L. E. Overman, G. Pairaudeau, *J. Am. Chem. Soc.* **1993**, *115*, 9293–9294.
6. M. J. Eichberg, R. L. Dorta, K. Lamottke, K. P. C. Vollhardt, *Org. Lett.* **2000**, 2, 2479–2481.
7. G. J. Bodwell, J. Li, *Angew. Chem., Int. Ed.* **2002**, *41*, 3261–3262. (*Angew. Chem.* **2002**, *114*, 3395–3396).
8. D. Solé, J. Bonjoch, S. García-Rubio, E. Peidró, J. Bosch, *Angew. Chem. Int. Ed. Engl.* **1999**, *38*, 395–397. (*Angew. Chem.* **1999**, *111*, 408–410).
9. M. Mori, M. Nakanishi, D. Kajishima, Y. Sato, *J. Am. Chem. Soc.* **2003**, *125*, 9801–9807.
10. Y. Kaburagi, H. Tokuyama, T. Fukuyama, *J. Am. Chem. Soc.* **2004**, *126*, 10246–10247.
11. T. Ohshima, Y. Xu, R. Takita, M. Shibasaki, *Tetrahedron* **2004**, *60*, 9569–9588.
12. M. V. King, J. L. DeVries, R. Pepinsky, *Acta Crystallogr., Sect. B.* **1952**, *5*, 437–440.
13. J. Schreiber, W. Leimgruber, M. Pesaro, P. Schudel, A. Eschenmoser, *Angew. Chem.* **1959**, *71*, 637–640.
14. J. Schreiber, W. Leimgruber, M. Pesaro, P. Schudel, T. Threlfall, A. Eschenmoser, *Helv. Chim. Acta* **1961**, *44*, 540–597.
15. T. Graening, H.-G. Schmalz, *Angew. Chem., Int. Ed.* **2004**, *43*, 3230–3256. (*Angew. Chem.* **2004**, *116*, 3292–3318).
16. R. B. Woodward, *The Harvey Lecture Series* **1963**, *59*, 31.
17. D. A. Evans, S. P. Tanis, D. J. Hart, *J. Am. Chem. Soc.* **1981**, *103*, 5813–5821.
18. T. Graening, W. Friedrichsen, J. Lex, H.-G. Schmalz, *Angew. Chem., Int. Ed.* **2002**, *41*, 1524–1526. (*Angew. Chem.* **2002**, *114*, 1594–1597).
19. T. Graening, V. Bette, J. Neudörfl, J. Lex, H.-G. Schmalz, *Org. Lett.* **2005**, 7, 4317–4320.
20. J. C. Lee, J. K. Cha, *Tetrahedron* **2000**, *56*, 10175–10184.
21. E. J. Corey, B. E. Roberts, *J. Am. Chem. Soc.* **1997**, *119*, 12425–12431.
22. J. Boukouvalas, Y.-X. Cheng, J. Robichaud, *J. Org. Chem.* **1998**, *63*, 228–229.
23. S. R. Magnuson, L. Sepp-Lorenzino, N. Rosen, S. J. Danishefsky, *J. Am. Chem. Soc.* **1998**, *120*, 1615–1616.
24. D. Demeke, C. J. Forsyth, *Org. Lett.* **2000**, 2, 3177–3179.
25. R. Paczkowski, C. Maichle-Mössmer, M. E. Maier, *Org. Lett.* **2000**, 2, 3967–3969.
26. M. E. Krafft, Y.-Y. Cheung, C. A. Juliano-Capucao, *Synthesis* **2000**, 1020–1026.
27. M. E. Krafft, Y.-Y. Cheung, K. A. Abboud, *J. Org. Chem.* **2001**, *66*, 7443–7448.
28. L. A. Paquette, J. Tae, M. P. Arrington, A. H. Sadoun, *J. Am. Chem. Soc.* **2000**, *122*, 2742–2748.
29. P. A. Wender, N. C. Ihle, C. R. D. Correia, *J. Am. Chem. Soc.* **1988**, *110*, 5904–5906.
30. J. Limanto, M. L. Snapper, *J. Am. Chem. Soc.* **2000**, *122*, 8071–8072.
31. S. M. Weinreb, *Chem. Rev.* **2006**, *106*, 2531–2549.
32. T. J. Greshock, R. L. Funk, *Org. Lett.* **2001**, *3*, 3511–3514.
33. P. Sun, C. Sun, S. M. Weinreb, *J. Org. Chem.* **2002**, *67*, 4337–4345.

34. S. M. Weinreb, *Acc. Chem. Res.* **2003**, *36*, 59–65.
35. H. Abe, S. Aoyagi, C. Kibayashi, *Angew. Chem., Int. Ed.* **2002**, *41*, 3017–3020. (*Angew. Chem.* **2002**, *114*, 3143–3146).
36. C. Kibayashi, S. Aoyagi, H. Abe, *Bull. Chem. Soc. Jpn.* **2003**, *76*, 2059–2074.
37. H. Abe, S. Aoyagi, C. Kibayashi, *J. Am. Chem. Soc.* **2005**, *127*, 1473–1480.
38. H. Mayr, A. R. Ofial, J. Sauer, B. Schmied, *Eur. J. Org. Chem.* **2000**, 2013–2020.
39. K. Takai, O. Fujimura, Y. Kataoka, K. Utimoto, *Tetrahedron Lett.* **1989**, *30*, 211–214.
40. Proposal by the R. W. Hoffmann Group, **2001**.
41. W. H. Pearson, Y. Ren, *J. Org. Chem.* **1999**, *64*, 688–689.
42. R. Hunter, P. Richards, *Synlett* **2003**, 271–273.
43. J. Liu, J. J. Swidorski, S. D. Peters, R. P. Hsung, *J. Org. Chem.* **2005**, *70*, 3898–3902.
44. P. Schär, P. Renaud, *Org. Lett.* **2006**, *8*, 1569–1571.

Chapter 12
Summary and Concluding Remarks

Abstract The golden rules of synthesis planning are collated, rules that transcend and supplement the contents of the individual chapters.

The considerations given in the previous chapters can be summarized in the following way as the Golden Rules of Synthesis Planning:

Focus on the skeleton of your target:
  Look for hidden symmetry.
  Make cuts to halve the target structure or the intermediates.
  Adjust cuts to create branches (skeleton oriented retrosynthesis).
  Adjust cuts to the functionality present (FG oriented retrosynthesis).
  Adjust cuts to enable generation of stereogenic centers.

Recognize patterns of stereogenic centers and check method oriented retrosynthesis.
Recognize patterns of stereogenic centers and check availability from the chiral pool.

Set functionality in your starting materials and intermediates to minimize refunctionalization steps (avoid oxidation and reduction steps).
Set the sequence of your skeleton forming steps:
  to maximize synchronous or tandem multibond forming operations,
  to achieve an exponential increase in complexity,
  to minimize the necessity of protecting groups.

Check the robustness of your plan,
  the compatibility of your protecting group pattern with changes in the plan and sequence of steps.

Guidelines for planning and executing syntheses have been delineated as well by others, for example, in reference [1]:

R.W. Hoffmann, *Elements of Synthesis Planning*,
DOI 10.1007/978-3-540-79220-8_12, © Springer-Verlag Berlin Heidelberg 2009

"(1) Redox reactions that do not form C-C bonds should be minimized, (2) the percentage of C-C bond-forming events within the total number of steps in a synthesis should be maximized, (3) disconnections should be made to maximize convergency, (4) the overall oxidation level of intermediates should linearly escalate during assembly of the molecular framework (except in cases where there is a strategic benefit such as an asymmetric reduction), (5) where possible, cascade (tandem) reactions should be designed and incorporated to elicit maximum structural change per step, (6) the innate reactivity of functional groups should be exploited so as to reduce the number of (or perhaps even eliminate) protecting groups, (7) effort should be spent on the invention of new methodology to facilitate the aforementioned criteria and to uncover new aspects of chemical reactivity, (8) if a target molecule is of natural origin, biomimetic pathways (either known or proposed) should be incorporated to the extent that they aid the above considerations."

The high level of congruence between the two paragraphs above signals consent among synthetic organic chemists as to the important goals to be reached in the design of syntheses. Some of the points mentioned above invite an additional short comment: Reactions that do not form skeletal bonds lengthen syntheses in an unnecessary fashion. These are, for the most part, protecting group management steps and redox refunctionalizations. A typical sequence which appears repeatedly in many published syntheses today is shown in Scheme 12.1:

$$R-COOMe \longrightarrow R \diagdown OH \longrightarrow R \diagdown O \longrightarrow R \diagup COOMe \longrightarrow R \diagup OH$$

Scheme 12.1 Common redox sequence in synthesis

This sequence encompasses three redox steps and just one skeletal bond-forming operation, giving testimony of our inability to do this in just a single operation. Efforts [2, 3] to shorten the reaction sequence of Scheme 12.1 are therefore highly welcome.

Moreover, redox reactions are likely to be a source of problematic side products and waste. Therefore, it would be preferable to perform synthesis throughout by a method wherein the oxidation level of the intermediates remains constant. Syntheses meeting this criterion are called "isohypsic" syntheses [4, 5], and efforts to attain this goal are encouraged. A totally isohypsic reaction scheme for a synthesis is, however, likely to be unrealistic. Rather, the ultimate goal along these lines is to attain "redox-economy." This means that a multistep synthesis should involve only those redox reactions that are strategic, i.e., that generate a stereogenic center or that set the final oxidation state of the functional groups in the target. In this manner all redox steps that are merely refunctionalizations would become dispensable.

Finally, adherence to biosynthesis patterns in planning a synthesis has the potential to lead to highly effective syntheses, inasmuch as biosynthesis relies on the inherent reactivity of functional groups and does not require any protecting groups. The synthesis of the secodaphniphylline skeleton [6] shown in Scheme 12.2 was inspired by a proposed biosynthetic pathway. This route turned out to be much shorter than a previous synthesis carried out following conventional wisdom.

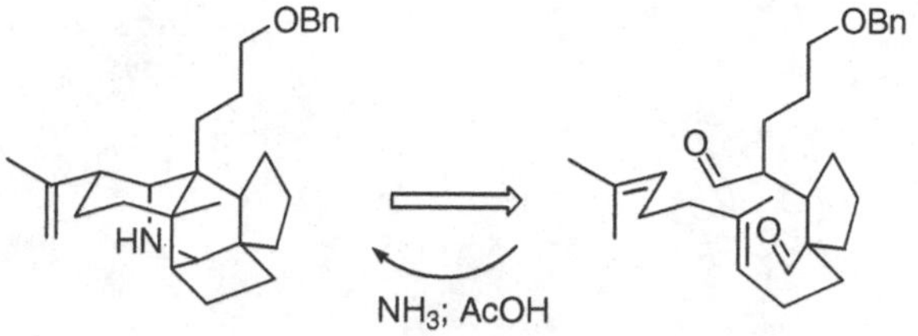

Scheme 12.2 Biosynthesis-inspired synthesis of the secodaphniphylline skeleton

To conclude the topic of synthesis planning, the best remark is that from C. H. Heathcock [6]:

"Although our approaches to problems have matured, we need even more mature strategies of synthesis. There is no reason that organic chemists should not be able to surpass nature's virtuosity in the synthesis of complex organic structures. In fact, we are still very far from this goal in most cases."

References

1. P. S. Baran, T. J. Maimone, J. M. Richter, *Nature* **2007**, *446*, 404–408.
2. J. Pospísil, I. Markó, *Org. Lett.* **2006**, *8*, 5983–5986.
3. R. R. Leleti, B. Hu, M. Prashad, O. Repic, *Tetrahedron Lett.* **2007**, *48*, 8505–8507.
4. J. B. Hendrickson, *J. Am. Chem. Soc.* **1971**, *93*, 6847–6854.
5. W. A. Smit, A. F. Bochkov, R. Caple, *Organic Synthesis: The Science behind the Art*, Royal Society of Chemistry, Cambridge, **1998**, pp. 98–118.
6. C. H. Heathcock, *Angew. Chem., Int. Ed. Engl.* **1992**, *31*, 665–681. (*Angew. Chem.* **1992**, *104*, 675–691).

Chapter 13
Solutions to Problems

In this section solutions are given to the problems delineated in the previous chapters. The answers given are by no means the only valid answers; rather they serve as examples of a suitable solution.

2.1 Brevicomins

The 1,2-diol unit allows a retrosynthetic disconnection between the two oxygen functionalities using an umpoled synthon, preferably on the side of the smaller fragment [1, 2, 3] (Scheme 13.1).

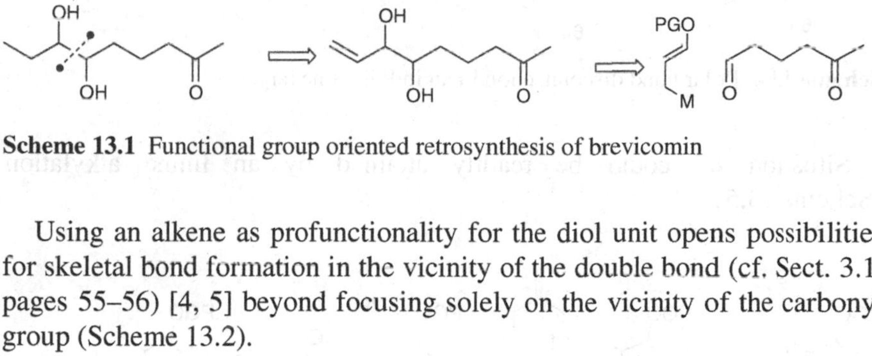

Scheme 13.1 Functional group oriented retrosynthesis of brevicomin

Using an alkene as profunctionality for the diol unit opens possibilities for skeletal bond formation in the vicinity of the double bond (cf. Sect. 3.1, pages 55–56) [4, 5] beyond focusing solely on the vicinity of the carbonyl group (Scheme 13.2).

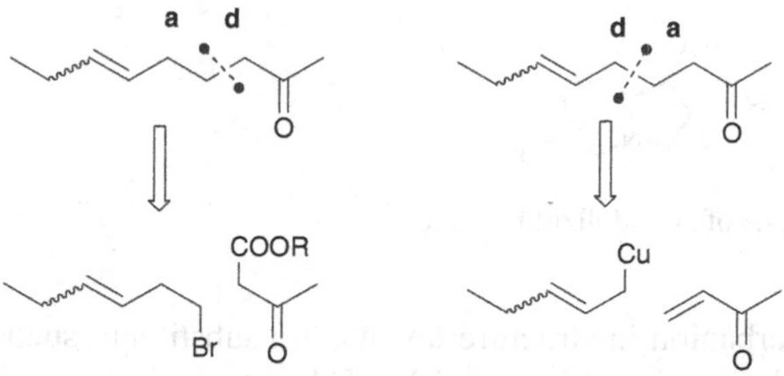

Scheme 13.2 Retrosynthesis of brevicomin considering a profunctionality

R.W. Hoffmann, *Elements of Synthesis Planning*,
DOI 10.1007/978-3-540-79220-8_13, © Springer-Verlag Berlin Heidelberg 2009

2.2 Indolizidines

The 1,3-distance of functional groups can be reached using synthons with natural polarity. The 1,2- and 1,4-distances indicated would require construction reactions using one umpoled synthon (Scheme 13.3).

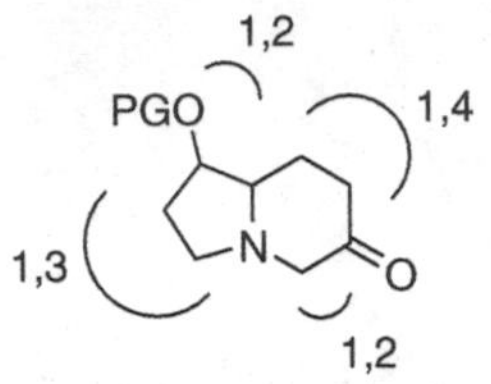

Scheme 13.3 Distance relationships between functional groups in an indolizidine target

Disconnection A in Scheme 2.65 suggests the following polarity patterns, each of which requires an umpoled synthon (Scheme 13.4).

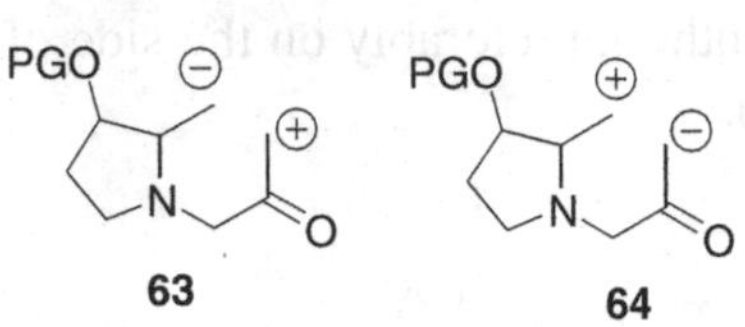

Scheme 13.4 Polar bond disconnection in an indolizidine target

Situation **63** could be readily attained by an imine alkylation (Scheme 13.5).

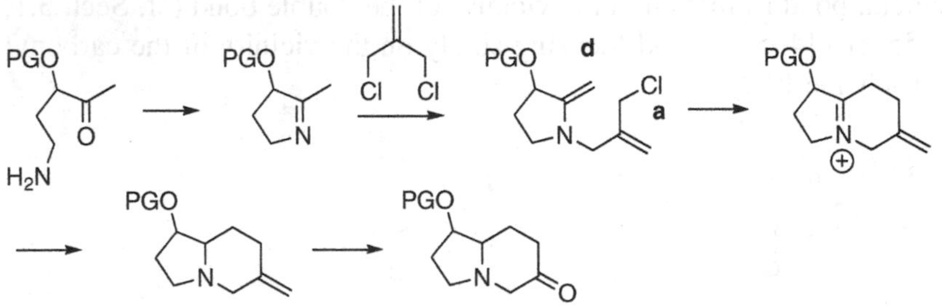

Scheme 13.5 Proposed synthesis of an indolizidine target

In order to stabilize a carbanion in structure **63** an extra substituent such as a sulfonyl group would be required (Scheme 13.6) [6].

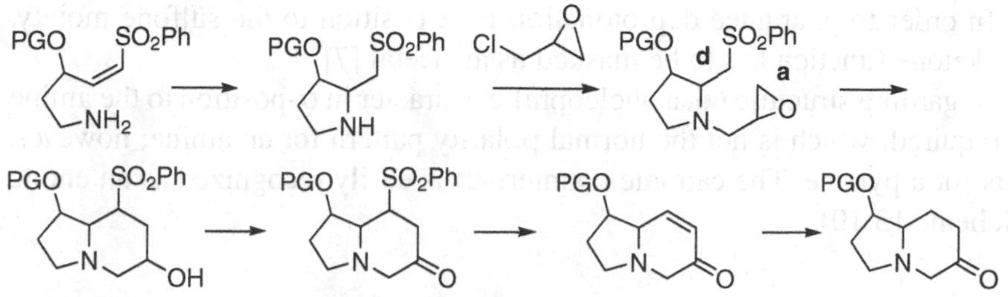

Scheme 13.6 Proposed synthesis of an indolizidine target

Bond A in Scheme 2.65 could also be considered to be formed in a (non-polar) ring closing metathesis reaction (Scheme 13.7).

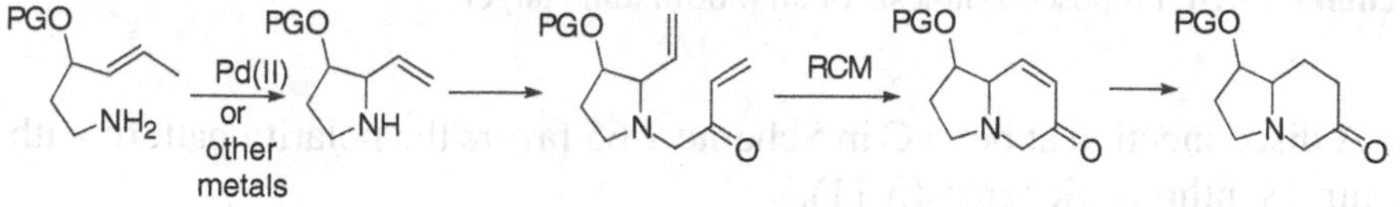

Scheme 13.7 Proposed synthesis of an indolizidine target

A disconnection at bond B in Scheme 2.65 leads to the following polarity patterns, again involving umpoled building units (Scheme 13.8).

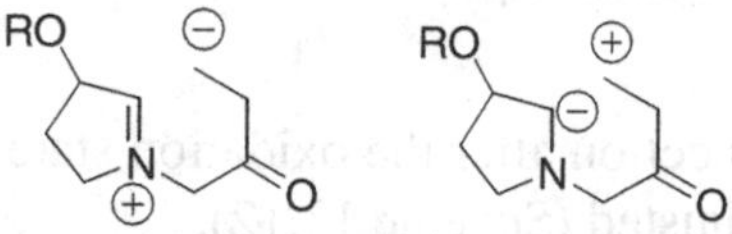

Scheme 13.8 Polar bond disconnection of an indolizidine target

In order to generate the carbanion in structure **65**, the use of a stabilizing substituent ($PhSO_2$) is indicated. The cationic part is easily identified as an iminium ion (Scheme 13.9).

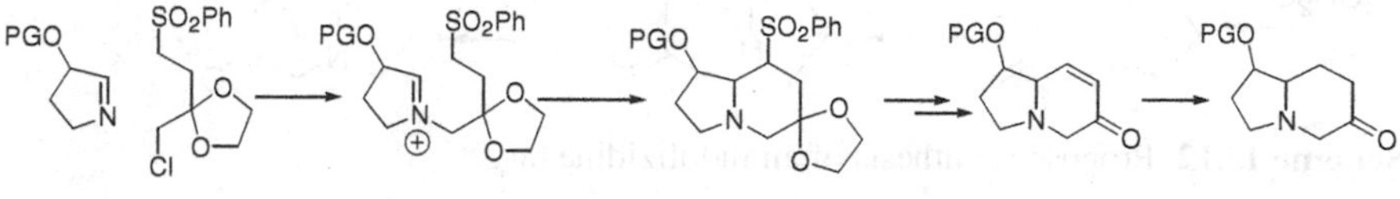

Scheme 13.9 Proposed synthesis of an indolizidine target

In order to guarantee deprotonation in α-position to the sulfone moiety, the ketone function has to be masked as an acetal [7].

Regarding structure **66**, a nucleophilic character in α-position to the amine is required, which is not the normal polarity pattern for an amine; however, it is for a pyrrole. The cationic counterpart is easily recognized as an enone (Scheme 13.10).

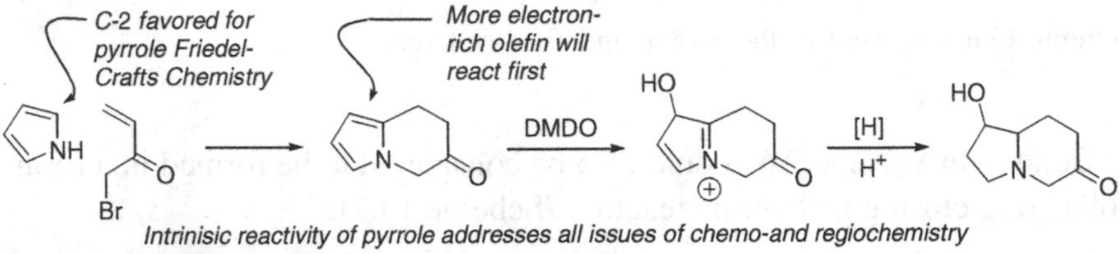

Intrinisic reactivity of pyrrole addresses all issues of chemo-and regiochemistry

Scheme 13.10 Proposed synthesis of an indolizidine target

A disconnection at bond C in Scheme 2.65 favors the polarity pattern with natural synthons (Scheme 13.11).

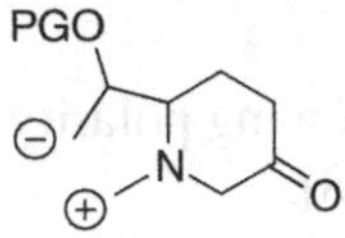

Scheme 13.11 Polar bond disconnection of an indolizidine target

This can be realized in a Mannich disconnection after the oxidation state of the alcohol function is correspondingly adjusted (Scheme 13.12).

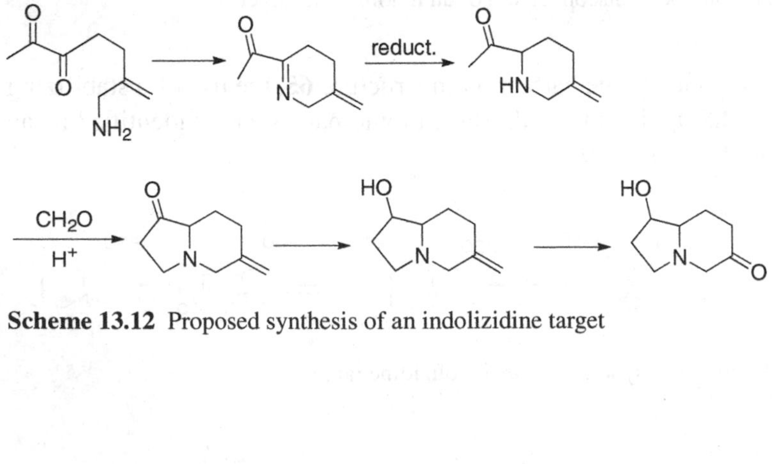

Scheme 13.12 Proposed synthesis of an indolizidine target

Masking of the final ketone as an alkene avoids problems in regioselective ketone reduction and avoids handling of an α-amino-ketone, with its high propensity to condense to dihydropyridazines.

3.1 Integerrinecic Acid

Making bond (3) (Scheme 3.25) is a general case of making α-hydroxy-acids. This is normally done via the cyanohydrin (Scheme 13.13).

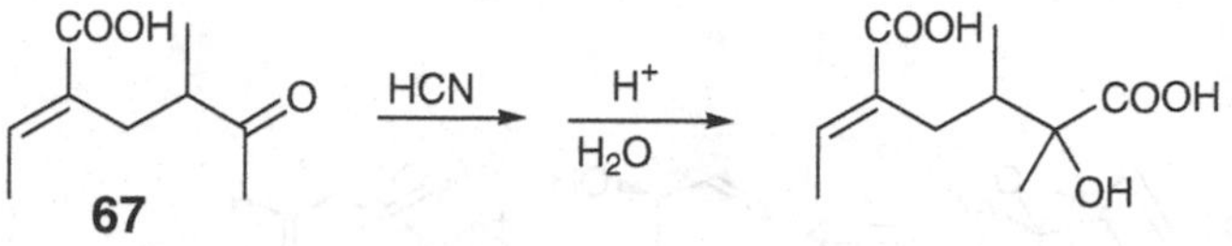

Scheme 13.13 Last step in a proposed synthesis of integerrinecic acid

This renders compound **67** a key intermediate, which can be accessed via formation of bond (2) (Scheme 13.14).

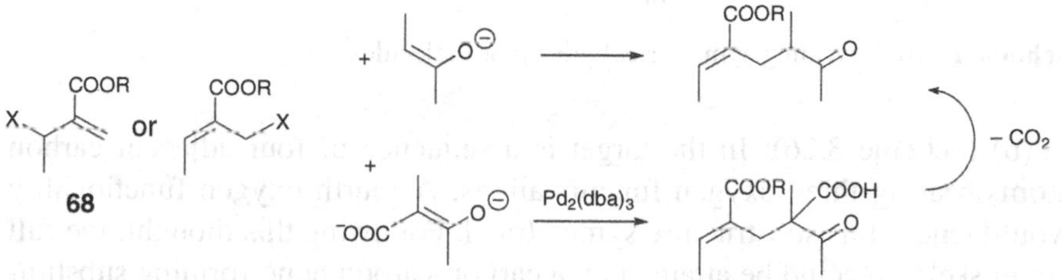

Scheme 13.14 Intermediate steps in a proposed synthesis of integerrinecic acid

Compound **68** in turn can be made by forming bond (1) (Scheme 3.25) using the Morita-Baylis-Hillman reaction (Scheme 13.15) [8].

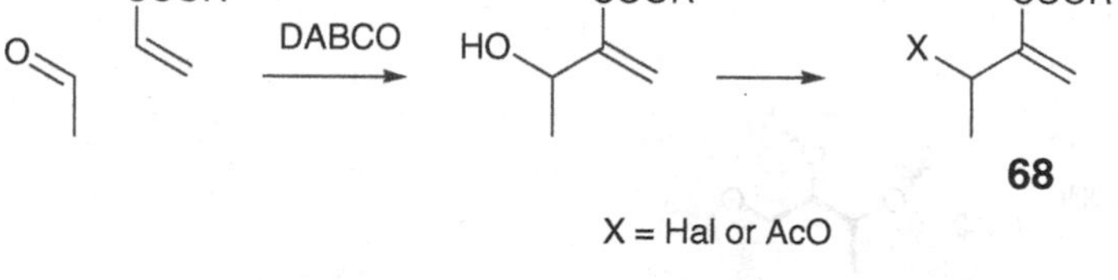

X = Hal or AcO

Scheme 13.15 Initial steps in a proposed synthesis of integerrinecic acid

3.2 Symmetrical Intermediates

(a) (Scheme 3.26): A symmetrical substructure is highlighted by bold bonds in the following scheme. A cut at the endocyclic double bond points to the possibility to enlist ring-closing metathesis in the forward synthesis. Disconnection of the acetal then leads to the c_2-symmetric dienediol [9] (Scheme 13.16). The latter can be obtained from D-mannitol. Likewise one could envision a symmetrical tartrate-derived dialdehyde as precursor. Here the task would be to effect two consecutive Wittig-reactions selectively (Scheme 13.16).

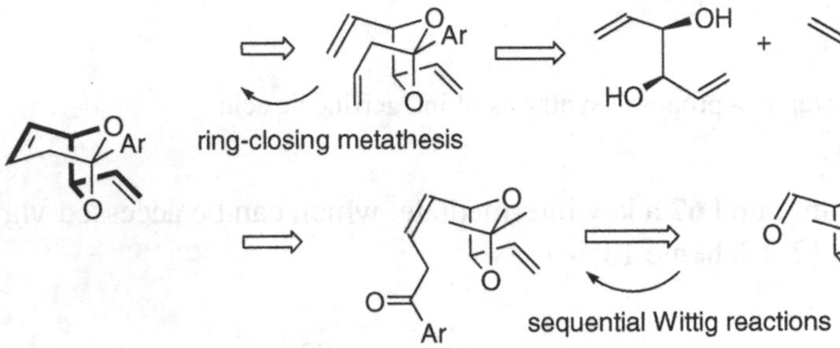

ring-closing metathesis

sequential Wittig reactions

Scheme 13.16 Revealing symmetrical precursor molecules

(b) (Scheme 3.26): In the target is a sequence of four adjacent carbon atoms bearing three oxygen functionalities. A fourth oxygen functionality would render this substructure symmetrical. Following this thought, the full target skeleton could be attained by a carbon-carbon bond forming substitution of that additional oxygen function. Along these lines one recognizes an epoxide as a potential substrate for such a substitution and after considering the stereochemistry of the epoxide opening one identifies the c_2-symmetric epoxidiol as starting material [10] (Scheme 3.17). The latter can be derived from tartaric acid. The c_2-symmetry of the epoxide renders both epoxide carbon atoms homotopic. It therefore doesn't matter at which one a nucleophilic acetylide attacks.

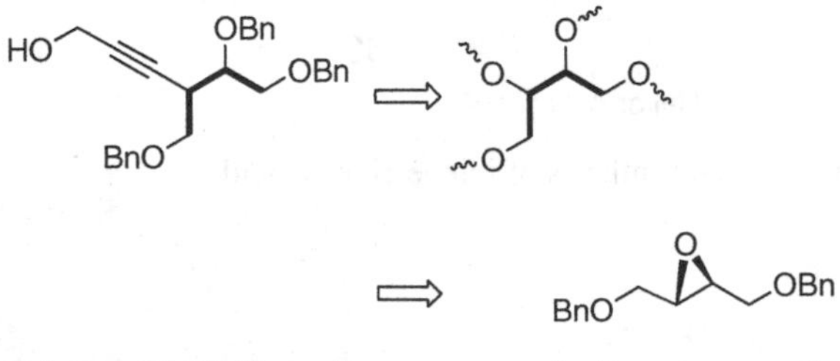

Scheme 13.17 Revealing a symmetrical precursor molecule

4.1 Polyoxamic Acid

When only the vic-diol unit of polyoxamic acid (Scheme 4.9) is considered, L-tartaric acid appears as a straightforward precursor for a building block oriented synthesis [11, 12] (Scheme 13.18). This leaves the task of addressing a stereoselective generation of the amine-bearing stereogenic center, preferably by substrate-based asymmetric induction (cf. Chap. 10).

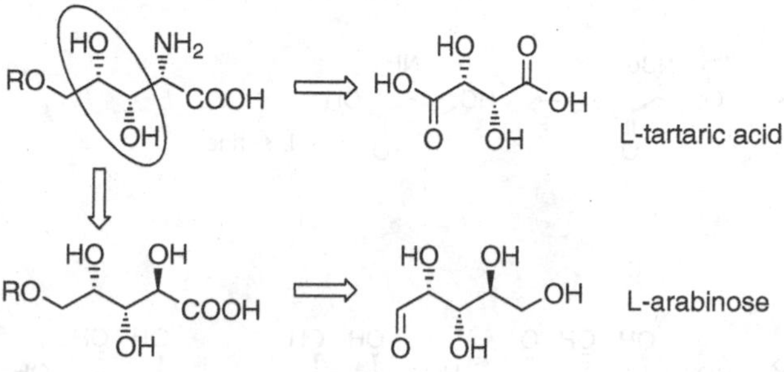

Scheme 13.18 Identifying precursor molecules for polyoxamic acid from the chiral pool

However, the amine-bearing stereogenic center could advantageously be generated from an alcohol function in a suitable precursor. Since this would likely involve an inversion of the configuration, the tetrahydroxy-pentanoic acid shown in Scheme 13.18 would be an attractive intermediate. Perusal of the list of sugar structures shown in Schemes 4.5 and 4.6 identifies L-arabinose as a readily available starting point [13].

An alternate way to polyoxamic acid is to start from a precursor with only a single stereogenic center, vinylglycine. Cross metathesis with allylic alcohol would assemble the molecular skeleton, leaving the stereoselective introduction of the two hydroxyl groups, a task to be accomplished by a Sharpless asymmetric dihydroxylation (Scheme 13.19).

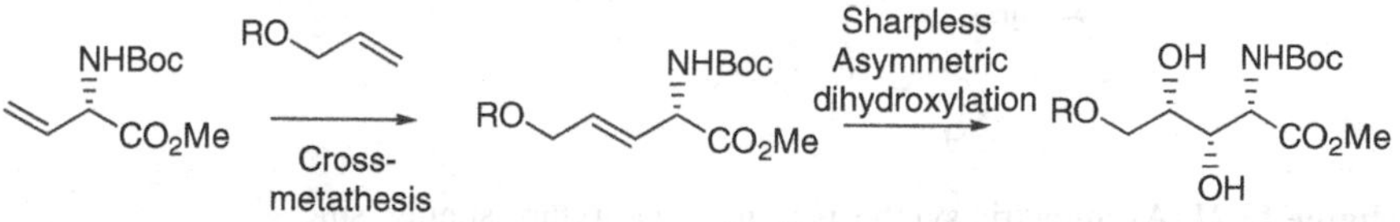

Scheme 13.19 Identifying vinylglycine as possible precursor for polyoxamic acid

4.2 D-Erythro-Sphingosine

The vicinal functionalities and the amine-bearing stereogenic center suggest L-serine as the most obvious precursor for a building-block oriented synthesis of D-erythro-sphingosine (Scheme 4.10). Yet this requires the construction of a second stereogenic center. When one envisions that the amine function could be generated by substitution with inversion from an alcohol, one is led to a vicinal trihydroxy-alkane as intermediate, and hence to search the sugars as precursors. This way one identifies D-galactose, D-xylose and D-arabinose as suitable starting materials for a synthesis of D-erythro-sphingosine [14] (Scheme 13.20).

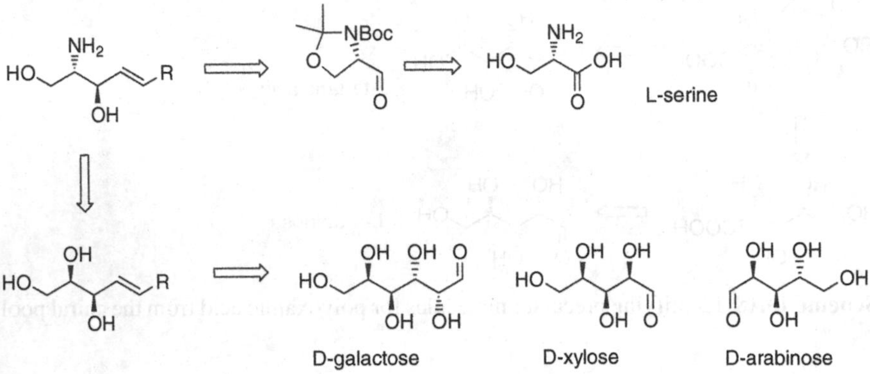

Scheme 13.20 Identifying precursor molecules for D-erythro-sphingosine from the chiral pool

The substitution pattern of D-erythro-sphingosine (Scheme 4.10) could also arise from an epoxy-alcohol. This suggests the Sharpless epoxidation as a suitable inroute [15] (Scheme 13.21).

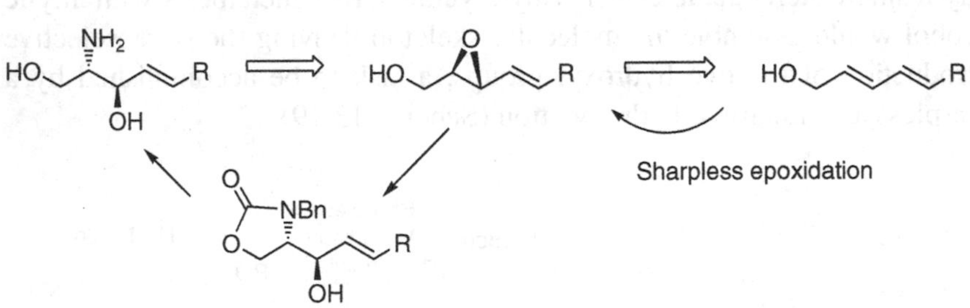

Scheme 13.21 Asymmetric synthesis route to D-erythro-sphingosine

5.1 Conjunctive Reagent

Formally, the simplest conjunctive reagent of the type shown in Scheme 5.9 would be "methylene" itself. That approach would require the simultaneous presence of the nucleophile R^- and the electrophile R^+ in solution. This is, however, not viable, as the latter may not be compatible with one another. Hence, a sequential coupling to the nucleophile and later to the electrophile, or vice versa, is indicated. For a carbene, however, addition of a nucleophile generates another nucleophile, which may likewise react with additional carbene leading to an undesired oligomerization (Scheme 13.22).

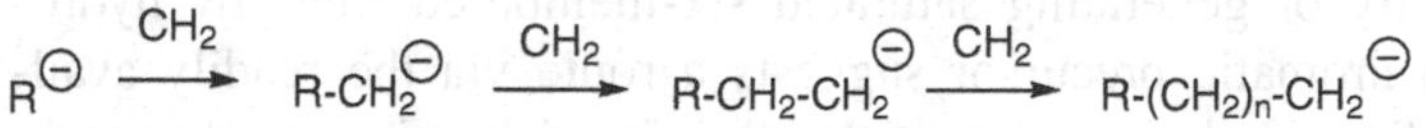

Scheme 13.22 Anionic oliogomerization of a carbene

The situation may be ameliorated by tuning of the reactivity of the "carbene," by a change to carbenoids as reagents. For example, ICH_2ZnI may be used as a conjunctive reagent [16], yet oligomerization still remains a problem (Scheme 13.23).

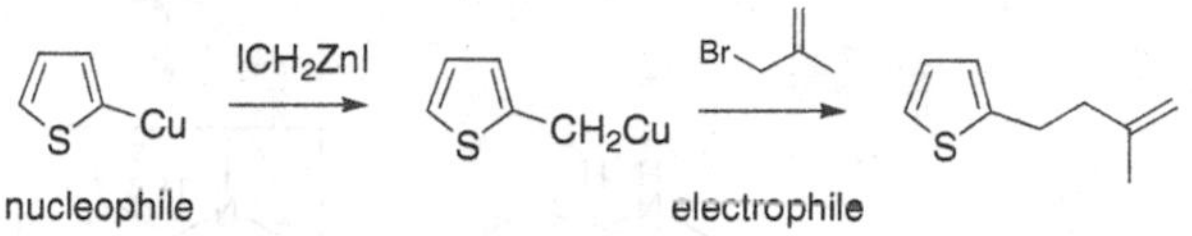

Scheme 13.23 Carbenoids allow for a stepwise homologation

It may therefore be advantageous to have a conjunctive reagent whose electrophilic and nucleophilic activity can be sequentially activated. An inconspicuous reagent in this context is dimethyl sulfoxide [17, 18, 19, 20, 21] (Scheme 13.24).

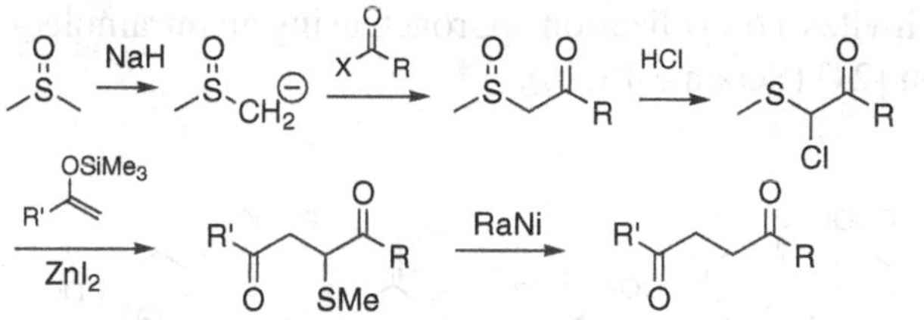

Scheme 13.24 Sequential generation of nucleophilic and electrophilic reactivity at the carbon atom of dimethyl sulfoxide

6.1 1,5-Diazadecalin

With bicyclization by reductive amination of 1,8-diamino-4,5-diketo-octane, there is no guarantee of forming the six-membered rings of

1,5-diaza-*cis*-decalin (Scheme 6.58) instead of the five-membered rings of a bis-pyrollidine.

Because of the availability of both L- and D-lysine, this compound constitutes a suitable starting point to allow access to either enantiomer of the product in a building block oriented approach, which is abbreviated in Scheme 13.25.

The possibility of generating saturated six-membered rings by hydrogenation of an aromatic precursor suggests a route via the readily available naphthyridine. Hydrogenation of the pyridine rings gives ready access to the racemic 1,5-diaza-*cis*-decalin, which can easily be resolved [22, 23] (Scheme 13.25).

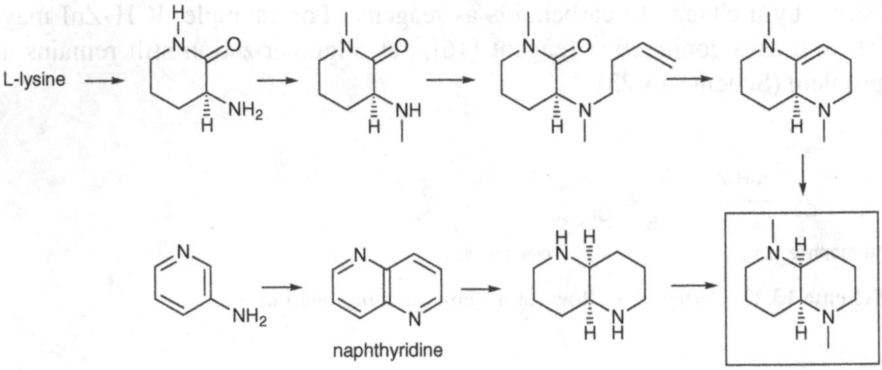

Scheme 13.25 Approaches to 1,5-diaza-*cis*-decalin

6.2 Dodecahydro-Anthracene

The double bond present in the central six-membered ring of the target structure (Scheme 6.59) invites a bicyclization approach using an intramolecular Diels-Alder addition [24] (Scheme 13.26).

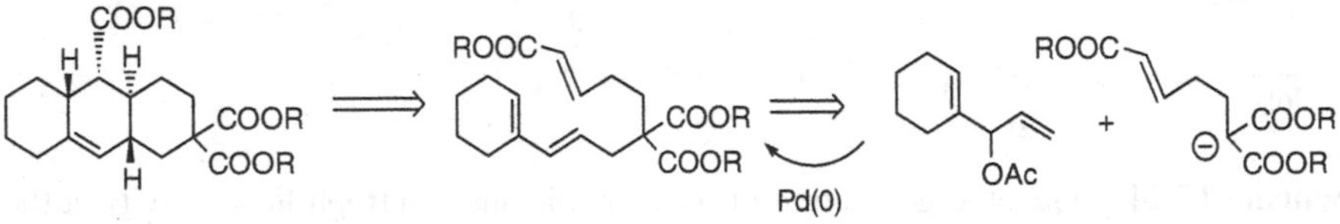

Scheme 13.26 Two-bond disconnection to generate a dodecahydro-anthracene skeleton

6.3 Twistane

The following four different bonds in twistane (Scheme 6.60) are available for a retrosynthetic cut [25] (Scheme 13.27).

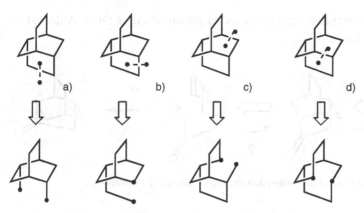

Scheme 13.27 Retrosynthetic cuts of twistane

Cleavage of the two-atom bridge (a) leads to a bridged precursor with two pending chains. Yet this most unfavorable solution has been used in a forward synthesis [26]. Cleavage of the two-atom bridge at (b) leads to a bridged precursor with one pending chain, an approach that allowed the first synthesis of twistane [27]. A similar situation results on cleavage of the one-atom bridge (c). The most significant retrosynthetic simplification results from the cleavage of the zero-atom bridge (d). This leads to a rather simple precursor system of two anellated six-membered rings.

The most highly bridged ring of twistane is marked in **69** (Scheme 13.28)

69

Scheme 13.28 Strategic bonds for the synthesis of twistane

defining strategic bonds. Cleavage of the zero-atom bridge as in d) is the top choice. A twistane synthesis according to this bond set has been realized [28] (Scheme 13.29).

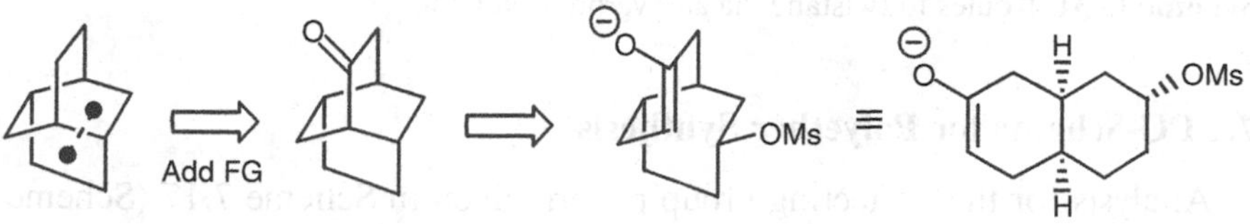

Scheme 13.29 The most advantageous route to twistane

A two-bond disconnection suggests an intramolecular Diels Alder addition (Scheme 13.30).

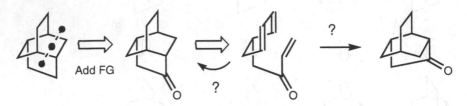

Scheme 13.30 Two-bond disconnection for the synthesis of twistane

However, there is a regioselectivity problem, which will favor the formation of a less strained isomer [25].

An add-bond strategy via an overbred skeleton offers the possibilities (a) and (b) shown in Scheme 13.31. Choice (a) suggests an intramolecular photocycloaddition between a cyclobutene and a cyclohexene—not really attractive. Choice (b) indicates a well precedented cyclopropanation reaction, that was indeed used [29] to realize a quick synthesis of twistane. However, this faced the problem of the regioselective cleavage of a distinct cyclopropane bond.

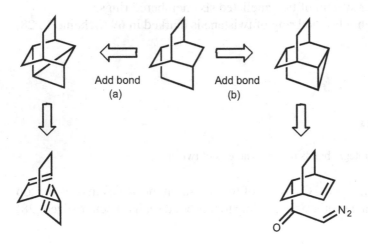

Scheme 13.31 Routes to twistane via an overbred skeleton

7.1 PG-Scheme for Polyether Synthesis

Analysis for the protecting group pattern given in Scheme 7.17 (Scheme 13.32):

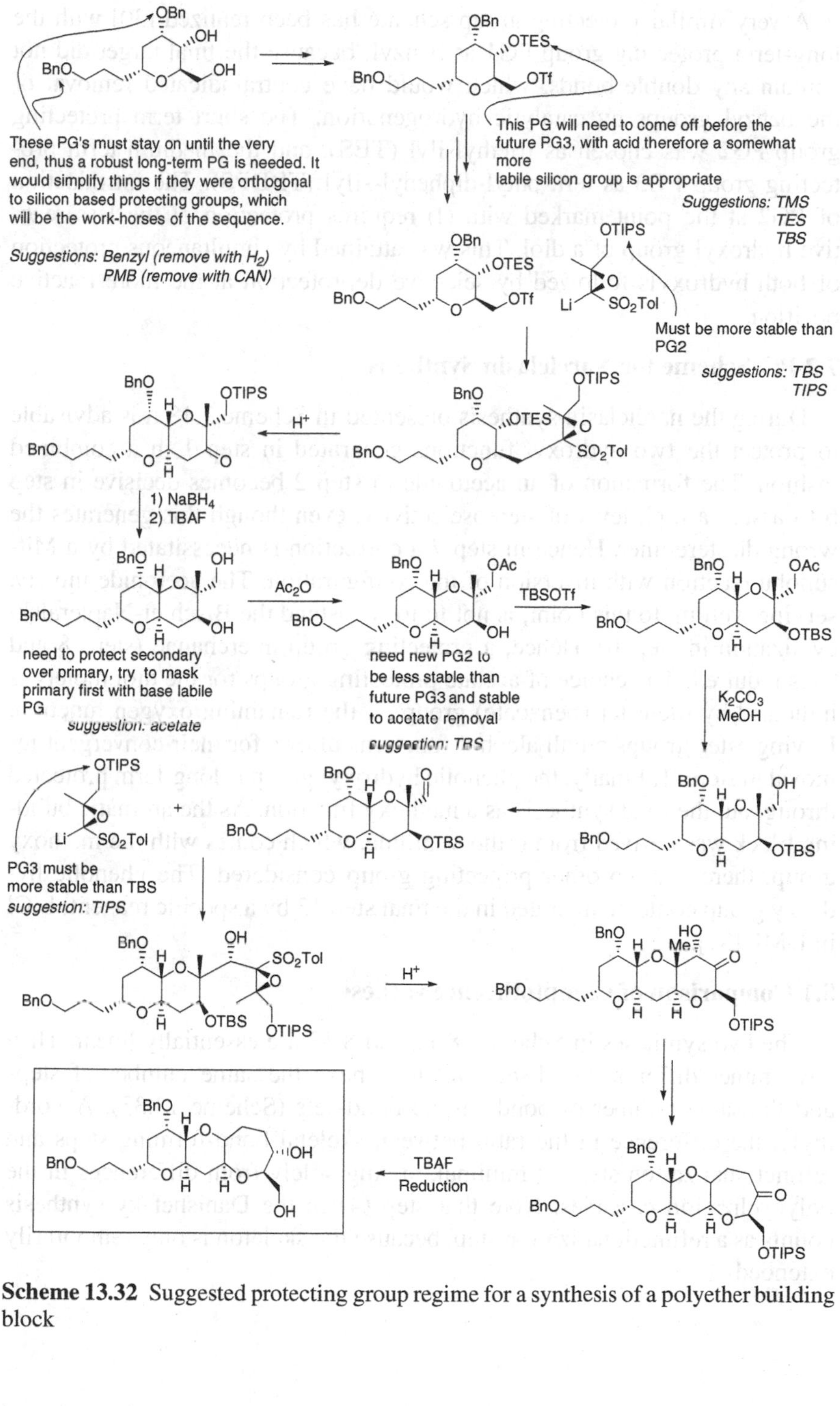
OBn
OH
BnO
OH
OTES
OTf
This PG will need to come off before the future PG3, with acid therefore a somewhat more labile silicon group is appropriate
Suggestions: TMS
TES
TBS
These PG's must stay on until the very end, thus a robust long-term PG is needed. It would simplify things if they were orthogonal to silicon based protecting groups, which will be the work-horses of the sequence.
Suggestions: Benzyl (remove with H2)
PMB (remove with CAN)
OTIPS
Li
SO2Tol
Must be more stable than PG2
suggestions: TBS
TIPS
H+
1) NaBH4
2) TBAF
OAc
Ac2O
TBSOTf
OTBS
need to protect secondary over primary, try to mask primary first with base labile PG
suggestion: acetate
need new PG2 to be less stable than future PG3 and stable to acetate removal
suggestion: TBS
K2CO3
MeOH
PG3 must be more stable than TBS
suggestion: TIPS
HO
Me
TBAF
Reduction

Scheme 13.32 Suggested protecting group regime for a synthesis of a polyether building block

A very similar protecting group scheme has been realized [30] with the long-term protecting group PG1 as benzyl, because the final target did not contain any double bonds, which would have contraindicated removal of the benzyl groups by catalytic hydrogenation. The short term protecting group PG2 was chosen as triethylsilyl (TES), and the medium term protecting group PG3 as tert.-butyl-diphenyl-silyl (TBDPS). The introduction of PG2 at the point marked with (!) requires protection of the less reactive hydroxyl group of a diol. This was attained by simultaneous protection of both hydroxyls followed by selective deprotection at the more reactive position.

7.2 PG-Scheme for Narciclasin Synthesis

During the narciclasin synthesis presented in Scheme 7.18 it is advisable to protect the two hydroxyl functions generated in step 1 in a combined fashion. The formation of an acetonide in step 2 becomes decisive in step 6 to assure a high level of stereoselectivity, even though this generates the wrong diastereomer. Hence in step 7 a correction is necessitated by a Mitsunobu reaction with inversion of the configuration. The acetonide moiety, serving well up to this point, is not fit to withstand the Bischler-Napieralski cyclization in step 10. Hence, a protecting group interchange (steps 8 and 9) is required. The choice of acetate protecting groups for the diol moiety is indicated by the ester (benzoate) group on the remaining oxygen function. Having ester groups on all alcohol functions allows for their convergent removal in step 11. Finally, the phenolic hydroxyl group is long-term protected throughout the total synthesis as a methoxy function. As the aromatic building block was derived from ortho-vanilline, which comes with the methoxy group, there was no other protecting group considered. The phenolic hydroxy group could be liberated in the final step 13 by a specific reagent, LiCl in DMF [31].

8.1 Comparison of Camptothecin Syntheses

The two syntheses in Schemes 8.12 and 8.13 are essentially linear. They have rather different bond-sets, yet they have the same number of steps and the same number of bonds in the bond-sets (Scheme 13.33). Accordingly, the difference in the ratio between skeletal bond-forming steps and refunctionalization steps is minimal, arising solely from differences in the polycyclization cascades. Note that step (4) in the Danishefsky synthesis counts as a refunctionalization step, because the skeleton is only temporarily extended.

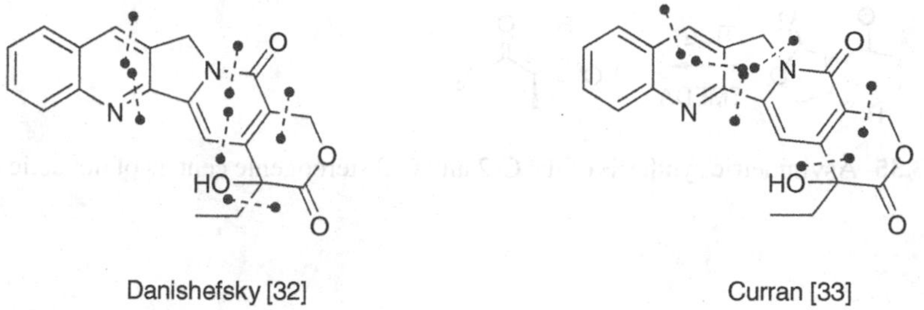

Danishefsky [32]   Curran [33]

Scheme 13.33 Bond-sets of Danishefsky's and Curran's camptothecin synthesis

Danishefsky's synthesis reaches a high level of complexity by the middle of the synthesis sequence (step (6)), whereas in the Curran synthesis high complexity is reached in the final step of the sequence. In terms of the robustness of the plan, the Danishefsky synthesis has substantial potential for variation. Curran's synthesis has two key steps, (3) and (8), on which failure or success depends. To have such a step as the last one of a synthesis is quite daring.

10.1 Nonactic Acid

For a review on the syntheses of nonactic acid (Scheme 10.23) see reference [34]. When considering a synthesis from the left end, the stereogenic center at C-8 is that of a terminal methyl-carbinol. These are most frequently derived from (R)- or (S)-propenoxide [35]. A start from 3-hydroxybutyrate [36, 37] would also provide the oxygen functionality for C-6. Malic acid has also been used to provide the stereogenic center at C-8 of nonactic acid [38] (Scheme 13.34).

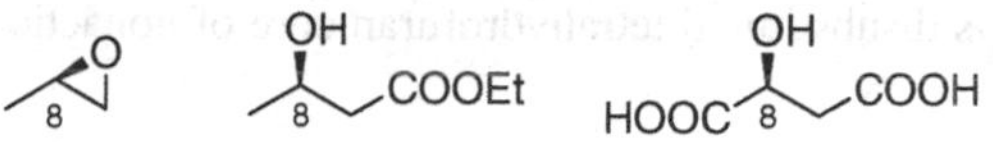

Scheme 13.34 Chiral pool precursors for the C-8 stereogenic center of nonactic acid

When a synthesis from the right end is considered, it is advantageous to generate both stereogenic centers at C-2 and C-3 in one stroke by an Evans-aldol reaction [35] (Scheme 13.35).

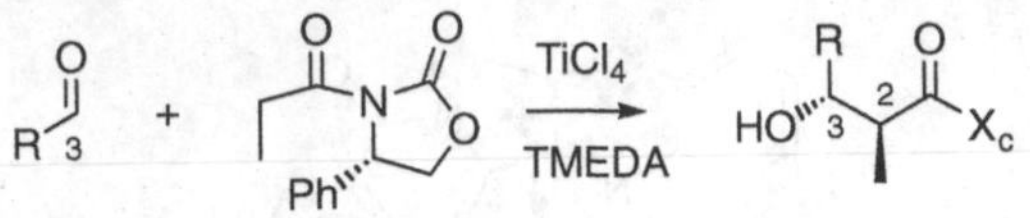

Scheme 13.35 Asymmetric synthesis of the C-2 and C-3 stereogenic centers of nonactic acid

For the relative configuration at C-3 and C-6, i.e., to establish the cis arrangment of the two side chains at the THF ring, a multitude of approaches are available [34] (Scheme 13.36). These include approaches utilizing a 1,4-asymmetric induction.

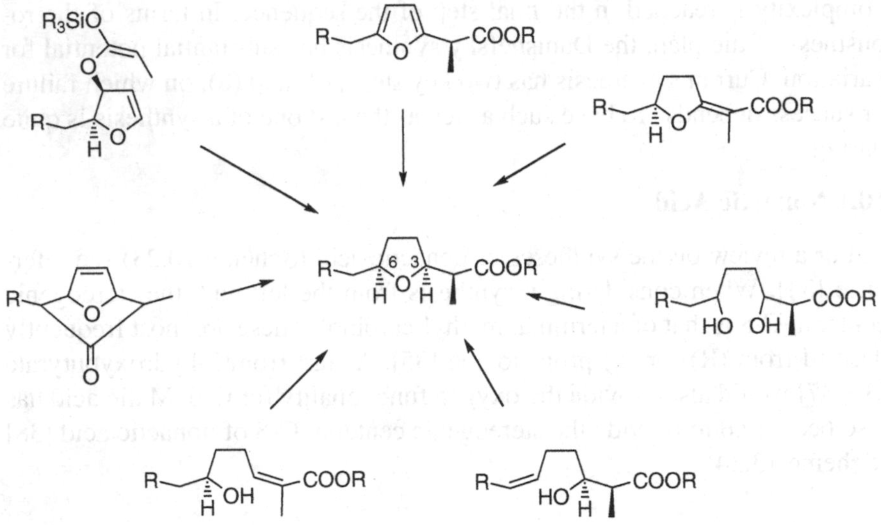

Scheme 13.36 Ways to generate the cis disubstituted tetrahydrofuran core of nonactic acid

The 1,3-distance of stereogenic centers at C-6 and C-8 is one that can be mastered by classical routes of asymmetric synthesis (Scheme 13.37).

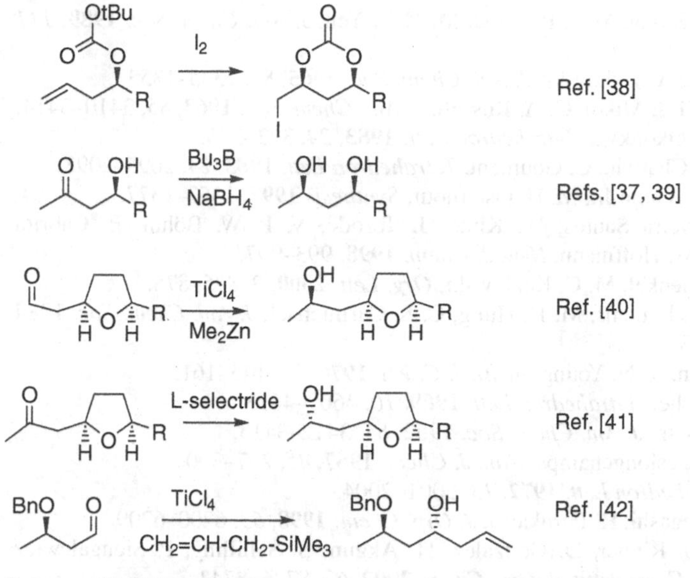

Scheme 13.37 Ways to set up the relative configuration at C-6/C-8 of nonactic acid

These various ways to exert stereocontrol may then be merged into an effective synthesis of nonactic acid.

References

1. R. W. Hoffmann, B. Kemper, *Tetrahedron Lett.* **1982**, *23*, 845–848.
2. Y. Yamamoto, Y. Saito, K. Maruyama, *Tetrahedron Lett.* **1982**, *23*, 4959–4962.
3. P. G. M. Wuts, S. S. Bigelow, *Synth. Commun.* **1982**, *12*, 779–785.
4. H. H. Wasserman, E. H. Barber, *J. Am. Chem. Soc.* **1969**, *91*, 3674–3675.
5. I. R. Trehan, J. Singh, K. Ajay, J. Kaur, G. L. Kad, *Indian J. Chem.* **1995**, 34*B*, 396–398.
6. J. C. Carretero, R. G. Arrayás, J. Org. Chem. **1998**, *63*, 2993–3005.
7. M. Julia, B. Badet, *Bull. Soc. Chim. Fr.* **1975**, 1363–1366.
8. D. Basavaiah, A. J. Rao, T. Satyanarayana, *Chem. Rev.* **2003**, *103*, 811–891.
9. S. D. Burke, E. A. Voight, *Org. Lett.* **2001**, *3*, 237–240.
10. K. C. Nicolaou, R. A. Daines, J. Uenishi, W. S. Li, D. P. Papahatjit, T. K. Chakraborty, *J. Am. Chem. Soc.* **1987**, *109*, 2205–2208.
11. F. Tabusa, T. Yamada, K. Suzuki, T. Mukaiyama, *Chem. Lett.* **1984**, 405–408.
12. A. K. Saksena, R. G. Lovey, V. M. Girijavallabhan, A. K. Ganguly, A. T. McPhail, *J. Org. Chem.* **1986**, *51*, 5024–5028.
13. A. Duréault, F. Carreaux, J. C. Depezay, *Tetrahedron Lett.* **1989**, *30*, 4527–4530.
14. P. M. Koskinen, A. M. P. Koskinen, *Synthesis* **1998**, 1075–1091.
15. R. Julina, T. Herzig, B. Bernet, A. Vasella, *Helv. Chim. Acta* **1986**, *69*, 368–373.

16. P. Knochel, N. Jeong, M. J. Rozema, M. C. P. Yeh, *J. Am. Chem. Soc.* **1989**, *111*, 6474–6476.
17. E. J. Corey, M. J. Chaykovsky, *J. Am. Chem. Soc.* **1965**, *87*, 1345–1353.
18. H.-D. Becker, G. J. Mikol, G. A. Russell, *J. Am. Chem. Soc.* **1963**, *85*, 3410–3414.
19. T. V. Lee, J. O. Okonkwo, *Tetrahedron Lett.* **1983**, *24*, 323–326.
20. L. Duhamel, J. Chauvin, C. Goument, *Tetrahedron Lett.* **1983**, *24*, 2095–2098.
21. A. Padwa, D. E. Gunn Jr., M. H. Osterhout, *Synthesis* **1997**, 1353–1377.
22. A. Gil de Oliveira Santos, W. Klute, J. Torode, V. P. W. Böhm, E. Cabrita, J. Runsink, R. W. Hoffmann, *New. J. Chem.* **1998**, 993–997.
23. X. Li, L. B. Schenkel, M. C. Kozlowski, *Org. Lett.* **2000**, *2*, 875–878.
24. B. M. Trost, M. Lautens, M.-H. Hung, C. S. Carmichael, *J. Am. Chem. Soc.* **1984**, *106*, 7641–7643.
25. D. P. G. Hannon, R. N. Young, *Austr. J. Chem.* **1976**, *29*, 145–161.
26. M. Tichý, J. Sicher, *Tetrahedron Lett.* **1969**, *10*, 4609–4613.
27. H. W. Whitlock Jr., *J. Am. Chem. Soc.* **1962**, *84*, 3412–3413.
28. J. Gauthier, P. Deslongchamps, *Can. J. Chem.* **1967**, *45*, 297–300.
29. M. Tichý, *Tetrahedron Lett.* **1972**, *13*, 2001–2004.
30. Y. Mori, K. Yaegashi, H. Furukawa, *J. Org. Chem.* **1998**, *63*, 6200–6209.
31. T. Hudlicky, U. Rinner, D. Gonzalez, H. Akgun, S. Schilling, P. Siengalewicz, T. A. Martinot, G. R. Pettit, *J. Org. Chem.* **2002**, *67*, 8726–8743.
32. W. Shen, C. A. Coburn, W. G. Bornmann, S. J. Danishefsky, *J. Org. Chem.* **1993**, *58*, 611–617.
33. D. P. Curran, S.-B. Ko, H. Josien, *Angew. Chem., Int. Ed. Engl.* **1995**, *34*, 2683–2684. (*Angew. Chem.* **1995**, *107*, 2948–2950).
34. I. Fleming, S. K. Ghosh *The Synthesis of Nonactic Acid* in *Studies in Natural Products Chemistry* (Ed.: Atta-ur-Rahman), Elsevier, Amsterdam, vol. 18, **1996**, pp. 229–268.
35. Y. Wu, Y.-P. Sun, *Org. Lett.* **2006**, *8*, 2831–2834.
36. E. Lee, S. J. Choi, *Org. Lett.* **1999**, *1*, 1127–1128.
37. P.-F. Deschenaux, A. Jacot-Guillarmod, *Helv. Chim. Acta* **1990**, *73*, 1861–1864.
38. P. A. Bartlett, J. D. Meadows, E. Ottow, *J. Am. Chem. Soc.* **1984**, *106*, 5304–5311.
39. I. R. Silverman, C. Edington, J. D. Elliott, W. S. Johnson, *J. Org. Chem.* **1987**, *52*, 180–183.
40. S. W. Baldwin, J. M. McIver, *J. Org. Chem.* **1987**, *52*, 320–322.
41. B. H. Kim, J. Y. Lee, *Tetrahedron Lett.* **1993**, *34*, 1609–1610.
42. T. Honda, H. Ishige, J. Araki, S. Akimoto, K. Hirayama, M. Tsubuki, *Tetrahedron* **1992**, *48*, 79–88.

Index